Geodesy for Geomatics and GIS Professionals

James A. Elithorp, Jr.
Dennis D. Findorff

Copley Custom Textbooks

An imprint of XanEdu Custom Publishing

Printed in the United States of America

ISBN 1-59399-087-1

Copley Custom Textbooks
An imprint of XanEdu Custom Publishing
300 North Zeeb Road
Ann Arbor, MI 48106
800-218-5971

Please visit our Web site at www.xanedu.com/copley

Preface

The life of this book began in the spring of 2000 in Little Rock, Arkansas at the Annual Convention of the American Congress on Surveying and Mapping. Over breakfast we discussed our common frustration that an introductory textbook in geodesy with a sound mathematical foundation was not available to our students. We agreed to work together to provide an undergraduate-level teaching textbook in geodesy for university students. The resulting text has been successfully used to teach the first course in geodesy at the Oregon Institute of Technology and at Troy State University for the past three years.

We are not unaware of the need that professional practitioners have for an understanding of fundamental geodetic theory. Rapid technological change has pushed the need for a working knowledge of geodesy from the halls of graduate school to the field and office of the geomatics professional. We therefore have prepared this introductory text with an eye toward those engaged in the daily challenges of professional practice in geomatics and geographic information systems. It is our hope that they, too, will benefit from our efforts.

The authors wish to acknowledge the donation of graphic material for the cover by Randy Fitch, PLS and Bret Elithorp, PLS of OBEC Consulting Engineers. We also wish to acknowledge the efforts of James A. Martin on the many figures and drawings contained in the book. Finally, we wish to express our appreciation for the influence of Earl F. Burkholder, and Dr. Boudewijn H.W. van Gelder as teachers in our own study of geodesy. Their love of the mathematics and science of geodesy was apparent to their students.

JAE
DDF

TABLE OF CONTENTS

CHAPTER ONE

INTRODUCTION

Depending on the accuracy required, and the purpose for which the geodetic computations are needed, we can employ as representing the earth's surface, either a plane, sphere, ellipsoid of revolution, triaxial ellipsoid, or geoid. The plane does not require any parameter. The sphere needs one parameter, the earth's radius (r). The ellipsoid of revolution requires two, the earth's equator radius (a) and the flattening (α) of the meridian. For the triaxial ellipsoid, four parameters are needed; (a) and (α), mentioned above, and in addition, the flattening of the equator, (α'), and the direction (λ₀) of the major axis of the equator. As for the geoid, it is not a mathematical surface, but depends on the irregular distribution of visible and invisible masses of matter near the earth's surface. Hence, it must be determined by observations, point by point.

If we have to map only a small region, as for instance the area of a city, the plane is sufficient. For computing the geodetic control points for larger areas, the curvature of the earth becomes appreciable, and we must use the ellipsoid or at least the sphere. The requirement for the triaxial ellipsoid arises more rarely in geodesy, but it has, if the triaxility truly exists, a large geophysical significance. The geoid, which theoretically is very important, will get, as we shall see, a great practical significance as well.
Heiskanen, 1951.

What is *geodesy*? This is probably the first question that comes to mind when you see the word. It is not a subject that often comes up in casual conversation. Perhaps the best way to start a discussion on geodesy is to review definitions proposed by renowned geodesists.

Definition of Geodesy

Geodesy literally means "*dividing the earth*" (Bomford, 1971). This definition has its derivation from the Greek word for geodesy (γη = earth, δαιω = I divide). A more comprehensive definition will be useful. F. R. Helmert, a leading geodesist of the 19th century, provided the classical definition: *Geodesy is the science of the measurement and mapping of the earth's surface* (Torge, 1991). This definition was probably adequate for Helmert's time, but with the advent of satellites and the multitude of observations (redundant measurements) made possible by them, this definition no longer suffices. The National Research Council of Canada proposed a more contemporary definition: *Geodesy is the discipline that deals with the measurement and representation of the earth, including its gravity field, in a three-dimensional time varying space* (Vanicek & Krakiwsky, 1982).

Finally, geodesy has also been described as a problem! *The problem of geodesy is to determine the figure and the external gravity field of the earth and of other celestial bodies as functions of time; as well as, to determine the mean earth ellipsoid from parameters observed on and exterior to the earth's surface* (Torge, 1991).

The last two definitions give you some idea of the breadth and depth of the science of geodesy. The inclusion of time in these definitions indicates that geodesy is a dynamic science due to the ever-changing nature of the earth's gravity field.

Definition of Geomatics

Geomatics is the science and technology of measuring, analyzing, interpreting, managing, displaying, distributing and using spatial data concerning both the Earth's physical features and the built environment. Geomatics encompasses a broad range of disciplines that can be brought together to create a detailed but understandable picture of the physical world and our place in it. These disciplines include surveying and mapping, geographic information systems (GIS), geodesy, global positioning systems (GPS), photogrammetry, and remote sensing.

Geomatics professionals recognize the importance of spatial data management brought about by the low cost of computing power, data storage, data collectors, and the routine need to transform data from one coordinate reference system to another. We shall use the term surveyor to denote the working professional in the field of geomatics throughout this text. In this context, a surveyor may be a photogrammetrist or a person specializing in subdivision design. The term 'land surveyor' is reserved for those persons specializing in boundary and cadastral surveys.

Relevance of Geodesy

You may ask, "So why do I need to learn about geodesy? I want to be a land surveyor, not a geodesist." Consider this: *Every land surveyor should be a geodesist, but not every geodesist is a land surveyor.* Now maybe this is a stretch (unless you look only at Helmert's definition of geodesy) since geodesy appears to deal more with the earth as a whole rather than the limited scope of a typical boundary survey, but today's surveyor must possess knowledge of the fundamentals of geodesy. Here are a few examples of the relevance of geodesy to contemporary land surveyors and professionals in the other sub disciplines of Geomatics.

Definition of Land Surveying

The statutory definition of the practice of land surveying in the State of Oregon includes: *Surveys (are) made for horizontal or vertical mapping control or geodetic control* (ORS 672.005(2)(c), 1995). Geodetic control must have something to do with geodesy! Many state statutes contain similar definitions that legally tie land surveying and geodesy.

Global Positioning System (GPS)

Geodesy is the backbone of GPS. Satellite orbits and positions, receiver positions, 3-D vectors, distances, and other values are described using coordinate systems defined in geodesy.

Geographic Information Systems (GIS)

The base layer for a GIS should be referenced to geodetic control. GIS projects that extend over large portions of the earth's surface have to deal with the earth's curvature. One solution is use of State Plane Coordinate Systems or the Universal Transverse Mercator (UTM) Coordinate System for these regional projects. A fundamental understanding of geodesy is necessary to effectively use these grid coordinate systems.

State Plane Coordinate System (SPCS)

State plane coordinate systems utilize an ellipsoid of revolution as a reference surface. Map projections are used to project points of interest on the curved ellipsoid surface to a flat plane or grid. SPCS are tools of great value to the surveyor, but are difficult to use in practice without a fundamental understanding of geodesy.

Coordinate Transformations

Surveyors routinely perform coordinate transformations, without even knowing it! Whenever you choose the "rotate", "translate" or "scale" functions in your coordinate geometry (COGO) program, you are performing 2-D coordinate transformations. Coordinate transformations are also integral to GPS. Coordinate transformations are an essential element of the study of geodesy.

Geodetic Boundary Value Problem

This problem describes the relationship between the earth's physical characteristics (gravity) and the geometric shape of the earth (ellipsoid). This relationship is investigated in geodesy.

Geodetic Surveying

We don't live on a flat earth, why do surveyors make this erroneous assumption? The obvious answer is that surveyors make this assumption either due to ignorance or for convenience in working projects of limited local extent. Surveyors must consider earth curvature when working on projects extending large distances (> 1 km). Geodesy allows surveyors to account for earth curvature.

Geodetic Datums

A geodetic datum is represented by a set of physical monuments on the earth's surface, published coordinates for these monuments, and a reference surface upon which the spatial relationships between the monuments are known. The provision of the geodetic datum is a public good that adds value to survey work. Knowledgeable clients require surveyors to tie their work to geodetic datums to give the spatial data greater value. The benefit of geodetic datums is compatibility between data sets. This compatibility is of great value to the development of GIS projects. The study of geodesy cannot neglect a treatment of the history and development of geodetic datums.

Employment Advantage

Unfortunately, many practicing surveyors know little or nothing about geodesy. With the continuing advance of technology, the need for persons possessing geodetic knowledge is becoming increasingly acute.

Our objective in providing this textbook is to help both the practicing surveyor and the geomatics student learn the fundamentals of geodesy.

Geodetic Subdivisions

It is convenient to subdivide the science of geodesy into narrower areas of study. We prefer the following subdivisions proposed by Burkholder (1987) and others:

Geometric Geodesy

Also known as *mathematical geodesy* or *terrestrial geodesy*, geometric geodesy deals with the size and shape of the earth as represented by geometric figures such as a sphere, biaxial ellipsoid of revolution, or triaxial ellipsoid.

Physical Geodesy

This branch of geodesy considers the earth's gravity field and equipotential surfaces, such as the geoid.

Satellite Geodesy

Also known as *extraterrestrial geodesy*, the relatively recent field of satellite geodesy is concerned with the study of geodesy using earth-orbiting satellites. Satellite geodesy comprises the observational and computational techniques that allow the solution of geodetic problems by the use of precise measurements to, from, or between artificial, mostly near-earth satellites (Seeber, 1993).

Geodetic Astronomy

This area of geodesy includes the earth's orbit, polar motion and determination of positions and azimuths on the earth's surface using celestial observations.

Plane Surveying

It is important to realize that the surveyor uses reference surfaces whether he or she is aware or knowledgeable of this fact. Thus, the term 'plane surveyor' is commonly used to describe one who uses a plane tangent to the earth's surface for computations. Use of the term, 'plane surveyor', may imply a lack of understanding of geodetic principles and a dependence on plane trigonometry for all calculations.

With the Land Ordinance of 1785, the U.S. Congress required that the rectangular survey system be established such that north south lines are oriented to true north and that east west lines are at right angles. This far-sighted legislation provided the framework for the description and conveyance of the public lands that remains a valuable asset to our economy today.

The Surveyor Generals that managed the work of the deputy surveyors on the ground had to face the problems generated by the placement of large rectangles on the curved surface of the earth. The problem is aggravated at the higher latitudes, and was addressed by the introduction of the correction line in Indiana in 1819 (McEntyre, 1985). As the U.S. Public Land System proceeded west, the correction lines were regularly spaced every 24 miles north or south of the baseline leading to the institutionalization of the quadrangle.

It is clear that in 1819 the implementation of the U.S. Public Land System raised serious problems for the 'plane surveyor'. Since the tangent plane as a reference surface does not require the use of parameters, it is possible for the plane surveyor to fail to realize that his/her reference surface is a plane by default. Since the plane surveyor does not think in terms of reference surfaces, it is especially difficult to identify the errors that can accumulate due to the use of a tangent plane to model the curved surface of the earth.

In this light, the U.S. Public Land Survey System provides a study of geodesy. With so bright an example in our history, it is surprising that many surveyors resist the study and application of geodetic principles. This limits their work to those small projects that can be accomplished on a tangent plane, or they may mistakenly apply the planar model to larger projects with resultant positional errors.

Reference Surfaces

Three important surfaces are useful in the study of geodesy: (1) the topographic surface; (2) the geometric reference surface; and, (3) the physical reference surface. The surveyor must understand the relationship between these three surfaces.

The topographic surface is the earth's surface upon which we live and work. Most often, the topographic surface does not correspond with either a chosen geometric reference surface or a physical reference surface.

The geometric reference surface is the mathematical reference system of choice in making geodetic calculations. This surface can be readily modeled mathematically. The sphere is a useful geometric reference surface. It is used to model the heavens (the celestial sphere) and as a model of the earth where great precision is not required. For example, one can calculate the distance between a place in Oregon and a place in Alabama using the sphere as the reference surface and spherical trigonometry for the calculations.

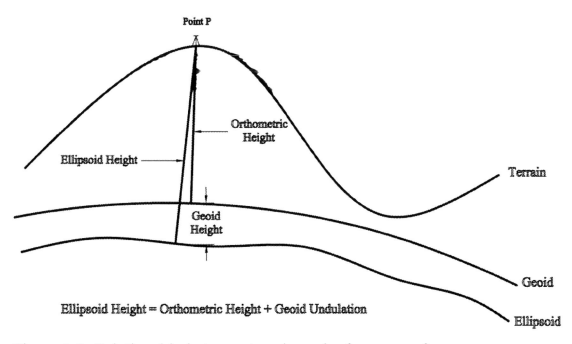

Figure 1.1 Relationship between terrain and reference surfaces.

The ellipsoid of revolution (or biaxial ellipsoid) is the geometric reference system of choice for calculating geodetic positions on the earth because it accounts for the flattening of the earth at the north and south poles due to centrifugal and gravitational accelerations. The ellipsoid of revolution is constructed by rotating an ellipse about an axis of rotation (minor axis). Determination of the earth's flattening and equatorial radius is necessary to size the ellipsoid model.

Mean sea level has long been used as a physical reference surface. It has been convenient because of the base level it represents for flowing waters. The geoid is now the preferred physical reference surface. The geoid closely corresponds with mean sea level but is defined by gravity potential. The geoid is a non-mathematical surface everywhere normal to the direction of gravity.

Consider a point P located on the earth's topographic surface. The distance from this point on the terrain to the ellipsoid surface is termed *ellipsoid height (h)*, the distance from this point on the terrain to the geoid is termed *orthometric height (H)*, and the distance between the geoid and the ellipsoid reference surface in terms of Point P is described as the *geoid height (N)*. Figure 1.1 demonstrates the relationship between these surfaces.

History of Geodesy (Short Version!)

The history of geodetic science can be traced back approximately 2200 years, perhaps farther. A very brief history is presented here. The Egyptian civilization dealt with the periodic flooding of the River Nile. While the periodic flooding provided a natural means of soil fertility, it also made the maintenance of land boundaries more difficult. While a detailed account of geodetic history is a fascinating study of man's progress in the sciences, it is not the focus of this textbook. Modern geodesy is dominated by satellite measurements and the development of physical geodesy. We present only a very brief history.

Spherical Earth Model

Long before Columbus and the debate over whether the earth was round or flat, an Egyptian by the name of Eratosthenes computed the radius of the presumed spherical earth using the simple relationship between arc length, radius and included angle: **s = r θ.** Eratosthenes was the first to use the scientific method not only to prove the world was "round" but also estimate the earth's radius. This may not seem like such an incredible feat, but consider that he did it in *220 B.C.!* Figure 1.2 illustrates Eratosthene's method.

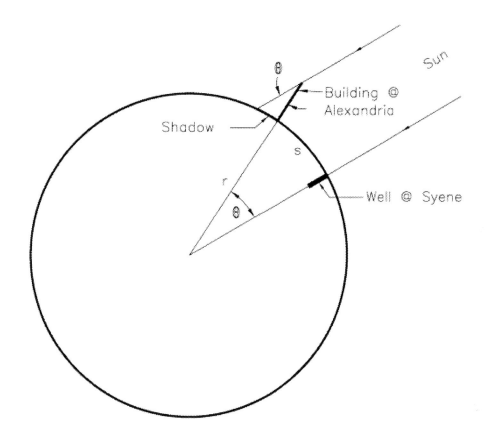

Figure 1.2 Reasoning of Eratosthenes.

Ellipsoidal Earth Model

During the late 17th century, scientists proposed an ellipsoidal model for the earth's shape rather than the commonly accepted spherical model. Newton and Huygens suggested an oblate ellipsoidal earth model, i.e., flattened at the poles, due to centrifugal acceleration. The Cassini Brothers, continuing the work of Picard under the direction of the Paris Academy, proposed a prolate ellipsoidal earth model, i.e., flattened at the equator. This hypothesis was based upon measurements made along a meridian and the use of celestial observations to determine latitude.

At this time in history, geodesists conducted numerous arc measurements along selected meridians to determine if the arc distance for a small change in latitude was the same along the meridian. Since the earth was expected to be flattened, geodesists sought to prove that the arc distance on the earth for the same change in latitude was different near the equator than near the poles. The famous debate between Newton and the Cassinis was resolved with two expeditions, one to Lapland (near the North Pole) and one to Peru (near the equator). The results supported the oblate model advanced by Newton and Huygens.

Geoid

During the late 19th century, Helmert coined the term *geoid* for the reference surface of physical geodesy that nearly coincides with mean sea level as proposed by Gauss earlier in the century. The geoid is the fundamental surface of physical geodesy to which orthometric heights are typically referenced.

Satellite Geodesy

The October 14, 1957 issue of LIFE Magazine reported that the U.S.S.R. had placed a 23-inch metal sphere into an orbit of approximately 560 miles above the earth, that this satellite was orbiting the earth every hour and 36 minutes broadcasting a radio signal. The headline read "**SOVIET SATELLITE SENDS U.S. INTO A TIZZY**." The signal was reported as: *An eerie intermittent croak—it sounded like a cricket with a cold—was picked up by radio receivers around the world last week.*

The launch of *Sputnik* in 1957 heralded the beginning of satellite geodesy. In 1993, the Global Positioning System (GPS) was declared operational. A constellation of 24 satellites propagates microwave signals at two different frequencies toward the surface of the earth. While the system was designed to provide navigation and timing capabilities for the U.S. Military, civilian uses immediately began to emerge. A major civilian use of GPS provides the determination of precise mark-to-mark distances (baselines) between points on the earth's surface for geodetic control purposes. GPS had an immediate and profound impact on the business of establishing and maintaining geodetic control datums in the United States. GPS has established satellite surveying as the province of all surveyors to establish geodetic control for boundary surveys and construction projects. GPS as a system involves reference surfaces, satellite orbits and positions, the propagation of electromagnetic energy through the earths' atmosphere, time, geodetic datums, coordinate transformations, and ellipsoid heights. The GPS equipment manufacturers have generally provided systems that are easy to use with very little or no training. The potential problem with such 'black box' technologies is that persons lacking the fundamental understanding of the technology or the expected results for each survey control project can regularly provide inaccurate results. The availability and capabilities of GPS systems has placed an even higher value on the study of the fundamentals of geodesy today.

CHAPTER TWO

LATITUDE AND LONGITUDE

When I'm playful I use the meridians of longitude and parallels of latitude for a seine, and drag the Atlantic Ocean for whales. **Mark Twain**

The shape of the earth is most nearly that of an oblate spheroid due to flattening at the poles. The ellipsoid of revolution is the more accurate mathematical model, but for many purposes the mathematical model of a sphere provides a sufficient geometric reference surface for calculations. We shall use the model of a sphere in this chapter to define and discuss the use of latitude and longitude to locate positions on the earth's surface.

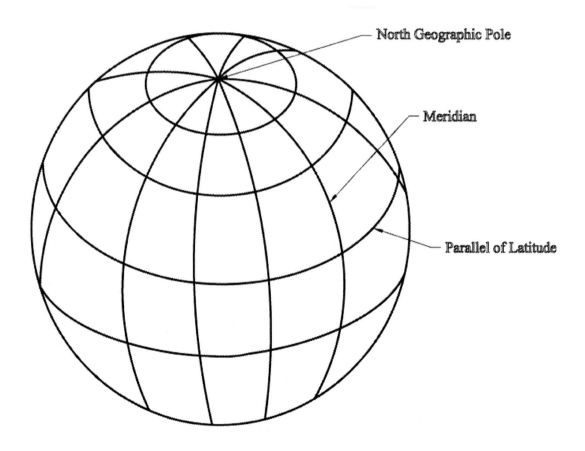

North Geographic Pole

Meridian

Parallel of Latitude

Figure 2.1 Lines of Longitude (Meridians) and Parallels of Latitude upon the Earth.

Latitude and longitude occupy an important history in the study of geodesy. This chapter will emphasize the fact that latitude is defined by the earth's natural coordinate system and that longitude is equivalent to time. Latitude and longitude are coordinates in the geographic coordinate reference system used to uniquely identify a place on the earth's surface.

Key to the understanding of this chapter is the fact that the earth constantly spins on a rotation axis. Just take a minute to reflect on the significance of the earth's rotation axis. The two points on the earth that are located at the ends of the rotation axis are the north and south geographic poles. At right angles to this rotation axis, at the point that divides the earth into two hemispheres, is the plane of the equator. The rotation axis and the plane of the equator provide the coordinate axes of the geographic coordinate system.

Definitions

Earth's Rotation Axis

The earth's rotation axis is the diameter upon which the earth rotates at a mean angular velocity (ω) that, with respect to our relatively stationary sun, gives a night and day time on the earth. Even though the rotation axis has been shown to move with time (polar motion) the axis serves as a natural line of reference.

Geographic Poles

The earth's poles are the points at which the rotation axis cuts the earth's topographic surface.

Great Circle

A great circle is a circle on the *surface of a sphere* formed by the intersection of a plane that passes through the center of the sphere.

Meridians

Meridians are great circles of the earth that pass through the north and south geographic poles. The plane of the meridian contains the earth's rotation axis. Each meridian is bisected into an upper and lower branch by the rotation axis of the earth. The upper branch is the portion of the meridian that passes through the particular point of interest on the earth's surface. There are an infinite number of meridians that may be considered. The earth can be cut through the plane of the meridian to form a meridian section. See figure 2.2 for illustration of the typical meridian section.

Meridians are north-south lines that can be used as reference lines from which directions can be reckoned. The direction found by standing on a meridian facing the north geographic pole is considered *true north*.

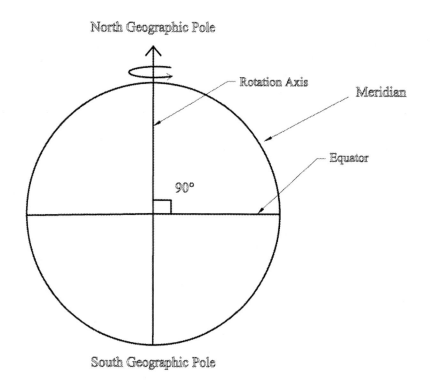

Figure 2.2 Typical Meridian Section—earth cut through rotation axis.

While meridians are great circles on a spherical model of the earth, they are ellipses on an ellipsoid of revolution model of the earth.

Small Circle

A small circle is a circle on the *surface of a sphere* formed by the intersection of a plane, not passing through the center of the sphere, with the sphere.

Parallels

Parallels are small circles on the sphere whose planes are parallel to the plane of the equator. The equator is the only parallel that is a great circle.

Latitude

The latitude of a place is its angular distance north or south of the equator measured along the meridian passing through that place. It is the angle at the earth's center subtended by the arc of the meridian from the equator to the place. Parallels are arcs of constant latitude.

Note: Latitude defined on the sphere will be denoted by the lower case Greek letter psi (ψ) pronounced "sigh". The reader should be aware that it is possible to define latitude in different ways on different geometric reference surfaces. We will use the symbol ψ for latitude when using the sphere as the geometric reference surface.

Problem 2.1: Troy, Alabama has a latitude of 31°48'02" North of the equator. How far, in miles, is Troy, Alabama from the equator?

It is convenient to use a sphere as a model of the earth. By convention, the radius of the spherical earth is taken as 6,371,000 meters. The formula for the arc of a circle (s = r θ) may be used to convert an angular distance θ, expressed in **radians**, into lineal distance (s) using the radius of the arc (r).

Refer to figure 2.3. The arc length from the equator to Troy may be expressed either as the included angle (ψ = latitude) or a lineal distance (s). Note that the radius of the spherical earth is R. We may now rewrite our formula as s = Rψ.

The first step is to convert the angle into decimal degrees.

ψ = 31° 48' 02" = 31.8006°

ψ = 31.8006° (π/180°) = 0.55502 radians

s = (6,371,000 meters) (0.55502 radians) = 3,536,060 meters or 3,536 kilometers

s = 3,536,060 meters (1 foot/0.3048 meters) (1 mile/5280 feet) =2,197 miles

This is an early valuable lesson in the study of geodesy. **Distances can be expressed as angles.** Angular distances have great value because the spatial relationship between two objects or points can be precisely expressed regardless of how far away the objects or points are from the origin or point of reference. Of course, if you wish to know the lineal distance between the two objects or points--their radial distance from the apex of the angle or origin must be determined.

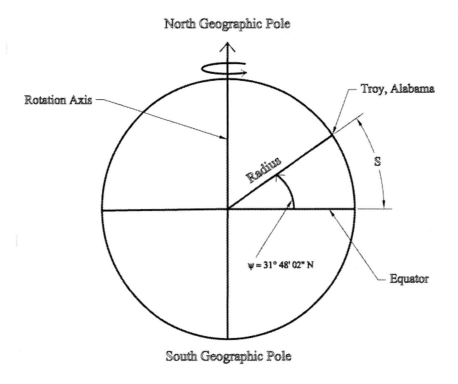

Figure 2.3 Latitude of Troy, Alabama shown in Meridian Section.

Longitude

The longitude of a place is its angular distance east or west of the prime meridian. The prime meridian is like any other meridian except it passes through a place deemed important by most countries of the world. This place is the Royal Observatory in Greenwich, England.

Longitude is designated by the Greek letter lambda (λ) pronounced "LAM-duh". Longitude is measured in the equatorial plane between meridians. The angle of longitude is formed by three intersection points: (1) the intersection of the prime meridian with the equator, (2) the intersection of the meridian through the point, for which the longitude is desired, with the equator, (3) and the intersection of the rotation axis with the equatorial plane. The angle of longitude can be reckoned either east or west. Be aware that geodesists consider east longitude to have a positive algebraic sign and west longitude to be negative. The angular distance from the meridian passing through

16

Greenwich, England and the meridian passing through Troy, Alabama is 85° 59' 56" W or 274° 00' 38" E. Stand on the north geographic pole looking down on the equator. A clockwise rotation gives west longitude and a counter-clockwise rotation gives east longitude.

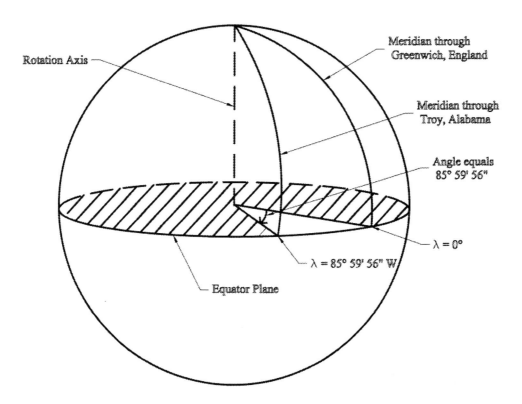

Figure 2.4 West Longitude of Troy, Alabama.

Lines of longitude are meridians. Consider a particular meridian that runs through Nebraska. It has a longitude of 100° W. It marks the accepted line between the arid western portion of the U.S. that requires irrigation to grow agricultural crops and the eastern portion that receives ample rainfall except in drought years.

Graticule

Parallels of latitude are curved east-west lines. Meridians of longitude are straight north-south lines. Meridians and parallels form a graticule or coordinate system that can uniquely locate a particular point on the earth's surface. Note that the lines of longitude and parallels of latitude meet at right angles.

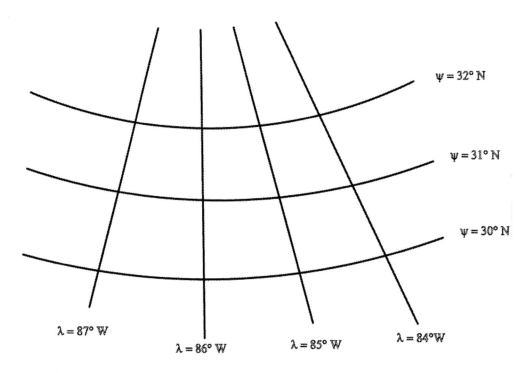

Figure 2.5 Graticule covering small portion of earth's surface.

Celestial Sphere

Let us step out into space with the concept of the celestial sphere. Note that while a spherical model of the earth approximates the real earth, the celestial sphere is pure imagination. Imagine a giant sphere of infinite radius with the sun, stars, and planets pushed out onto the surface of the sphere. The earth is just a dot at the center. The earth's north and south geographic poles can be projected outward along the rotation axis to become the celestial north and south poles. Similarly, the earth's equator is projected outward to become the celestial equator or equinoctial.

Zenith

The zenith is the point directly overhead. Consider an instrument carefully leveled up over a point on the earth's surface. The vertical axis of the instrument can be projected upward and outward to a unique point on the celestial sphere. A plane normal to the plumb line at this point on the earth forms a tangent plane. This tangent plane can also be termed a 'horizon plane' and can be projected out to the celestial sphere.

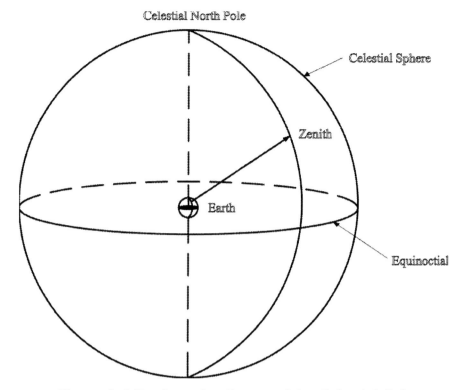

Figure 2.6 Earth at the Center of the Celestial Sphere.

Polaris as the Extended Pole

Very near the celestial North Pole is a star known as Polaris. You can find Polaris by turning a vertical angle equal to your latitude from your horizon plane northward. Two proofs can be offered for this fact:

Proof 1: Consider the sketch in figure 2.7. An observer has set a total station at point P on the earth. The tangent plane normal to this point P is the horizon plane. The angle POE is the observer's latitude by definition. The angle POH plus the angle POE is equal to 90 degrees because the equator is normal to the earth's rotation axis. Therefore Angle POH is equal to (90°-ψ). Consider angle OPH in triangle POH. By definition of the tangent plane (horizon) this angle has to be 90°. Therefore since the sum of the angles in a triangle is 180°, then angle PHO has to be the latitude (ψ). Since alternate angles formed by two intersecting straight lines are equal, then the vertical angle from your horizon plane to the extended pole or Polaris is your latitude.

A key to understanding the sketch in figure 2.7 is the realization that the distance between P and O is not to scale. If the sketch were drawn to scale,

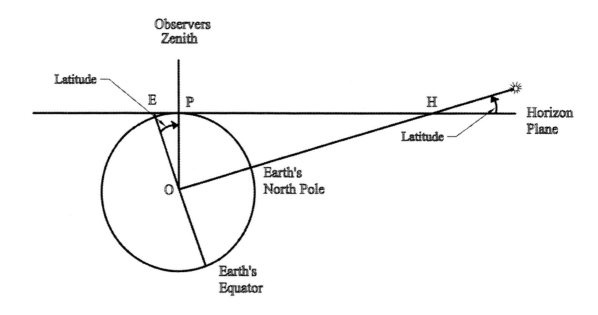

Figure 2.7 Sketch for Proof 1--Polaris is near the extended North Pole on the Celestial Sphere.

the infinite distance to Polaris on the celestial sphere would by proportions shrink distance OP to an infinitesimal segment. A drawing to scale might look more like the sketch in figure 2.8.

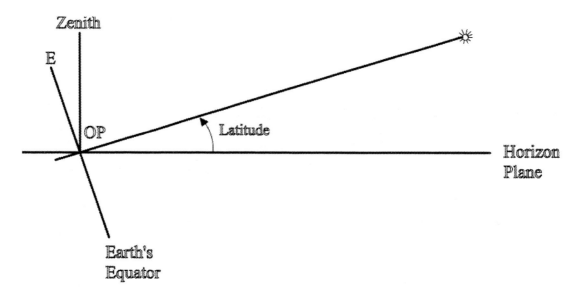

Figure 2.8 Field Sketch of Figure 2.7 showing proper proportion between distance to Polaris and radius of the Earth.

Proof 2: Consider the sketch in figure 2.9. Angle ZOE is your latitude by definition. Since Polaris is very near the Celestial North Pole and since the rotation axis of the earth is perpendicular to the equator by construction, then Angle POZ is (90° - ψ). Since the horizon plane is perpendicular to your zenith and the rotation axis is perpendicular to the equator, then Angle POH = Angle ZOE = ψ.

Angle ZOE = ψ.

ψ + Angle POZ = 90° and Angle POH + Angle POZ = 90°.

Therefore: ψ + Angle POZ = Angle POH + Angle POZ

Angle POZ can be subtracted from each side of the equation

ψ = Angle POH which is the vertical angle from the horizon plane up to Polaris.

There are several methods for finding the latitude of a place using geodetic astronomy, however a study of these procedures is not the subject of this text. The reader is referred to Dutton, 1943. However, it is important to realize that the rotation axis of the earth and the placement of the equator normal to the rotation axis provide a natural set of axes for the determination of latitude.

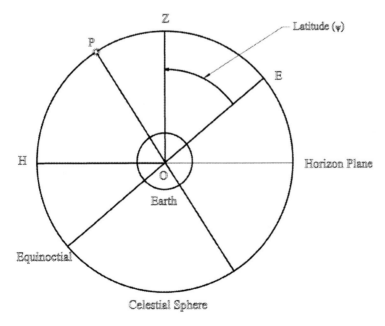

Figure 2.9 Sketch for Proof 2—Earth surrounded by Celestial Sphere.

Time

Early mariners could determine their latitude quite easily, but longitude being referenced in an arbitrary fashion remained a significant problem. Designating the meridian through Greenwich, England as 0° is arbitrary in the sense that any other place would have served just as well. In those days, ships could sail north and south within sight of land and know where they were, but once out of sight of land only latitude could be determined. This led to many shipwrecks involving captains who misjudged the east west distance to the coasts of Europe upon returning from sea. A good example of this problem was seen in the return of British ships from the open seas trying to find the English Channel, but instead being shipwrecked on the coast of Europe.

The solution of this problem was to realize that longitude is not only an angular distance, but it is also time. The earth can be assumed to rotate on its axis once every 24 hours. This allows the sun to transit or pass over your meridian once each day at noon. Since the sun cannot transit over two different meridians at the same time; then there must be a time difference between when the sun transits the prime meridian at Greenwich and any other meridian.

This difference can be quantified because 24 hours = 360°:

$$1^h = 15° \qquad\qquad 1° = 4^m$$
$$1^m = 15' \qquad\qquad 1' = 4^s$$
$$1^s = 15'' \qquad\qquad 1'' = 0.0667^s$$

Problem 2.2: A sailor sailed west from Greenwich, England carrying a precise clock set at Greenwich time. That is, when the sun crossed over the Prime Meridian, the time was 12:00 noon. Several days later on the open seas, the sailor decided to determine his longitude. The sailor read the time off his Greenwich clock when the sun crossed his local meridian as 15^h 18^m 55^s. What is his longitude?

Note that the sailor sailed west from Greenwich, England. From the vantage of the celestial North Pole, West is clockwise from the Prime Meridian.

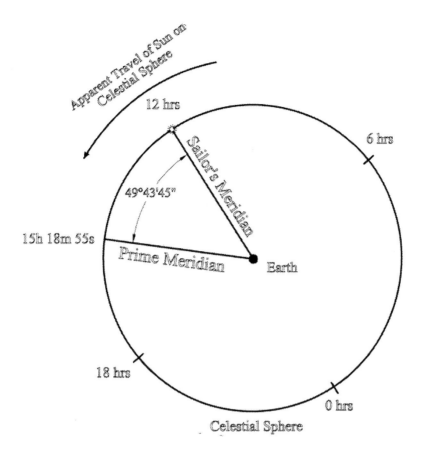

Figure 2.10 Earth Fixed—Apparent movement of Sun from west to east on Celestial Sphere.

The sun crosses local meridians to the west of the Prime Meridian at an earlier time. Using the sketch in figure 2.10 the solution follows:

$$15^h\ 18^m\ 55^s$$
$$\underline{-12^h\ 00^m\ 00^s}$$
$$3^h\ 18^m\ 55^s \Rightarrow 31.315278^h$$
$$31.315278^h(15°/1^h) = 49.729167° \text{ or } 49°\ 43'\ 45'' \text{ West Longitude.}$$

Students at this point may attempt to integrate their personal experience of seeing the sun rise in the East and set in the West. Figure 2.11 shows the apparent movement of the sun from East to West to a person standing on the surface of the rotating earth facing north. Let us try to relate this sketch to the one shown in figure 2.10. The model of the celestial sphere in figure 2.10 assumes a stationary earth. Therefore, the movement caused by the rotation of the earth is transferred to the sun on the surface of the celestial sphere. It

23

should be no surprise that the apparent movement of the sun has to be from West to East (opposite to that shown in figure 2.11) to duplicate the same rotational effect.

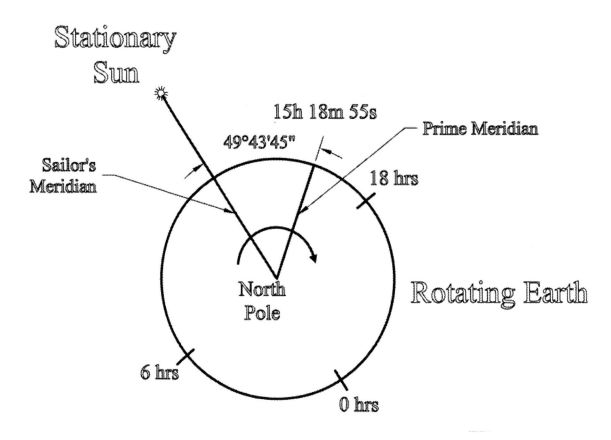

Figure 2.11 Earth Rotates under Stationary Sun (Sun appears to move from east to west).

The solution to the longitude problem was to develop a precise clock that could be carried on shipboard. Regulators (clocks that worked on the principle of a pendulum) were available that could provide relatively precise time for this application, but were adversely affected by the motion of ships at sea. History records that the English Parliament passed the Longitude Act of 1714 that offered a substantial financial incentive for the invention of a precise clock that could work on shipboard. John Harrison invented the first of four chronometers in 1736 that met the requirements. He experienced great difficulty, but finally collected his prize money--the equivalent of millions of dollars today.

Time is a complex subject affected both by natural phenomena and the needs of civilization. No attempt is made to duplicate the fundamental material about time and survey astronomy found in reference textbooks on surveying. Material on time is presented here to emphasize the connection between longitude and time.

The time interval of one apparent revolution of the sun about the earth is called an apparent or true solar day. Two important facts have to be acknowledged at this point in the discussion: (1) the sun actually moves very little with respect to the earth. The sun appears to move about the earth because the earth spins on its axis. (2) The earth turns approximately 360° with respect to the sun every 24 hours. Since the earth travels an elliptical orbit about the sun while it is spinning, every complete revolution or day can vary by a maximum of ± 16 minutes of 24 hours.

It was found convenient to invent a fictitious mean sun that travels about the earth at a uniform length of day. This time system is called mean time = civil time = universal time (UT). It is possible to compute the difference between mean (civil) and true (apparent) time using the equation of time.

Equation of time = true solar time – mean solar time.

Most calculations in geodesy require the use of a worldwide time zone. Such a time zone is Greenwich Mean Time (GMT) or Universal Time (UT). In this time zone, the day begins and ends on the Prime Meridian through Greenwich, England.

Universal Time (UT) is not sufficiently accurate for applications of survey and geodetic astronomy. UT1 is Universal Time corrected for polar motion and variations in the earth's rotational velocity and should be used by the surveyor in the conduct of astronomic observations. UT1 is broadcast in the United States by the National Institute of Standards and Technology and is received by short-wave radio receivers on frequencies of 2.5, 5, 10, and 15 MHz. UT1 may also be received via telephone at 303-499-7111.

The broadcast time actually given by human voice is UTC or Coordinated Universal Time. UTC is a time scale provided by a set of atomic clocks independent of the earth's rotation. A correction (DUT) is provided by a set of double clicks that immediately follow each minute tone of the broadcast signal. Each double click is a correction of 0.1 second and is positive for the first seven double ticks. Starting with the ninth second each double click is negative. UT1 is obtained by applying the formula:

$$UT1 = UTC + DUT$$

The United States was divided into standard time zones in 1883. This action was required because, theoretically, there exists an infinite number of meridians in the United States each with its own local time. To facilitate trade and commerce (namely the safety of railroad travel), it was convenient to select the local mean time of a standard longitude within each zone and apply that time to the whole standard time zone.

The standard longitude of the Eastern Standard Time Zone is 75° West; the Central Standard Time Zone is 90° West; the Mountain Standard Time Zone is 105° West; and the Pacific Standard Time zone is 120° West. The time in each Standard Zone differs from neighboring zones by 1 hour because the standard longitudes are 15° apart.

Problem 2.3: What is the time in the Central Standard Time Zone at 00^h Universal Time (UT)?

We remember that the West Longitude of the standard longitude of the Central Standard Time Zone is 90°. Since $1^h = 15°$ then:

$90°/(15°/1^h) = 6^h$

Furthermore, we know that west of the Prime Meridian is the previous day.

$24^h - 6^h = 18^h$ or 6:00 p.m. of the previous day.

Our last topic of this chapter is Daylight Savings Time (DST). It is also called "summer time." It is a method of artificially expanding the daylight hours with respect to the traditional workday. Clocks are set forward one hour in late March or early April and are set back one hour in late September or in early October. Not all States participate in Daylight Savings Time.

The State of Alabama does participate. The Daylight Savings Time of a time zone equals the Standard Time of the next time zone to the east. During Daylight Saving Time Period the Daylight Savings Time in the Central Time Zone is 5 hours earlier than Universal or Greenwich Mean Time because the clocks were set ahead one hour. In Problem 2.3, the answer is 7:00 p.m. of the previous day under Central Daylight Time.

Study Questions

1. What physical characteristic about the earth made it possible for man to define latitude as a coordinate?
2. Explain with a sketch how it is possible to find Polaris in the daytime if you know your latitude at a place of observation.
3. What does the term 'arbitrary' mean in terms of longitude?
4. What physical characteristic about the earth and our sun provide meaning to the statement that "longitude is time"?
5. If Greenwich Civil Time is known at a particular instant to be $0^h 34^m 17^s$ a.m., what time is it in the Central Time Zone? Note the meridian from which time is reckoned in the Central Time Zone has a longitude of 90° 00' 00"W.
6. The Local True (Apparent) Time at a place is $7^h 40^m 13^s$ p.m. The Longitude of the place is 85° 59' 29"W. What is Greenwich True (Apparent) Time at this instant?
7. If the longitude of a place in Klamath Falls, Oregon is 121° 47' 09", what Pacific Standard Time will show on your watch when the sun transits this meridian?
8. Define the term "graticule." What fundamental differences can you find between a graticule of the Klamath Falls, Oregon area and a USGS Topographic Quad Map of the same area?
9. At 1600 hours in the Pacific Standard Time Zone, what is Greenwich Mean Time?
10. What parallel of latitude is a great circle? Why?

CHAPTER THREE

GEOMETRY OF THE SPHERE

The Earth was small, light blue, and so touchingly alone, our home that must be defended like a holy relic. The earth was absolutely round. I believe I never knew what the word round meant until I saw Earth from space.
      Aleksei Leonov,   U.S.S.R. (Soviet Astronaut)

We know that the earth is a nearly spherical planet upon which are superimposed the surface irregularities created by land and sea, highland and lowland, mountains and valleys. However these topographic irregularities represent little more than a roughening of the surface. Since the radius of the earth is 6371 km and since the major relief features do not rise more than 9 km above or fall more 11 km below sea level, they are relatively less important than, say, the seam on a cricket ball or the indentations on the surface of a golf ball.
      D.H.Maling, 1992.

It is useful to begin with a simplification of the earth model as a sphere. While not the preferred model, the sphere does allow for close approximations of numerous quantities of interest to the surveyor. The sphere is a simple model requiring one parameter—that of its radius (R).

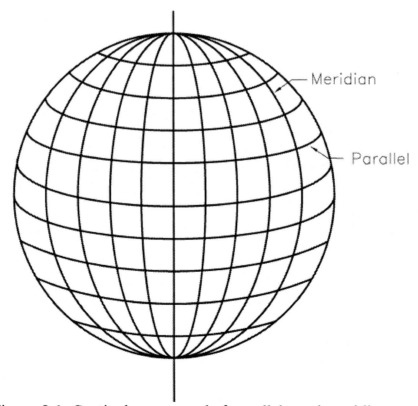

Figure 3.1 Graticule composed of parallels and meridians.

Basic Concepts

A geometric reference surface used to represent the earth may be chosen as a sphere having a radius (R) of 6371 km. It is advantageous to use a mathematical figure, such as the sphere, to represent the earth rather than the geoid that cannot be represented by a simple equation. The sphere may be subdivided using a *graticule* composed of *parallels* and *meridians*. Figure 3.1 shows a sphere with graticule.

A *great circle* is formed by passing a plane through the sphere so that the resulting great circle section contains the center of the sphere. A *small circle* is formed by any other intersection of a plane and a sphere (excepting a tangent intersection). Accordingly all meridians and the equator are great circles. All parallel circles (excepting the equator) are small circles. All parallel circles are parallel to the equator plane. Problem 3.1 requires computation of the lineal distance between two points located on a parallel of latitude.

Problem 3.1: Given point A and B on the parallel circle at a latitude of 42° 15' 16". Point A has a longitude of 121° 47' 09" W and Point B has a longitude of 119° 36' 29" W. Compute the parallel circle distance between A and B on the sphere of radius 6,371,000 meters.

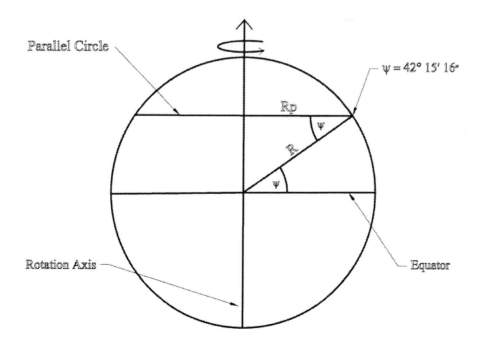

Figure 3.2 Radius of Parallel Circle.

The first step in the solution of this problem is to compute the radius of the parallel circle (R_p). Consider the meridian section of the sphere in figure 3.2.

$$\cos \psi = \frac{R_p}{R} \qquad so \ R_p = R \cos \psi$$

$$\therefore R_p = 6{,}371{,}000 \cos 42°15'16"$$

$$R_p = 4{,}715{,}597 \ meters$$

The second step is to compute the lineal distance between Points A and B using the familiar formula s = r θ. Consider the parallel section of the sphere in figure 3.3.

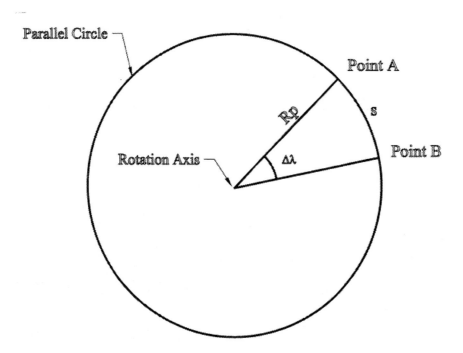

Figure 3.3 Looking down from North Pole on Parallel Circle.

The angle subtended at the center of the parallel by the arc between Point A and Point B is the difference in longitude.

Longitude of Point A ⇒ 121° 47' 09"W
Longitude of Point B ⇒ 119° 36' 29"W
2° 10' 40"

30

2° 10' 40" ⇒   2.1777777778° (π/180°) = 0.0380093925991 radians

s = 4,715,597.39258 meters (0.0380093925991) = 179,237 meters

Spherical Coordinates

The position of a point on or near the surface of a sphere may be made with respect to conveniently defined coordinate systems. Since more than one coordinate system may be used, it is useful to develop transformations between the systems. Figure 3.4 shows the relationship between curvilinear and Cartesian coordinate systems for the sphere.

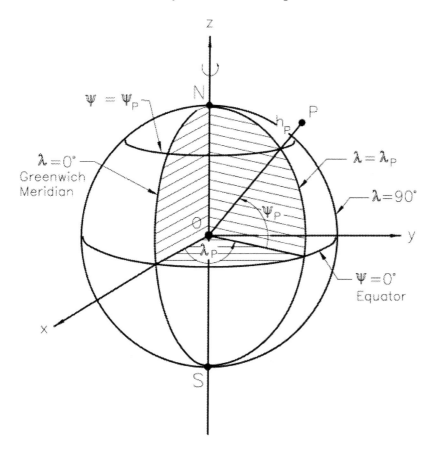

Figure 3.4 Coordinate Systems on the Sphere.

Curvilinear Coordinates (Geographic Coordinates)

A point on the surface of the sphere may be defined using a *curvilinear coordinate* pair: *latitude* and *longitude*. Latitude (ψ) is measured in a meridian plane along a meridional great circle from the equatorial plane to the north or south. North latitudes are assigned positive values while south latitudes are negative. Longitude (λ) is reckoned from the Prime Meridian and is measured in the equatorial plane. Longitude reckoned from the Prime Meridian counterclockwise or east is designated by convention, positive, in computations. Westerly measured longitudes are negative in such computations..

Points lying above or below the surface of the sphere may be further defined by the use of their height (h_s) above (positive) or below (negative) the surface of the sphere. Thus, any point may be described using a curvilinear coordinate triplet (λ, ψ, h_s).

Cartesian (Rectangular) Coordinates

A point position may also be expressed in terms of (rectangular) *Cartesian coordinates*. It is necessary to define a *right-handed orthogonal* set of coordinate axes before expressing a point position as an x, y, z triplet. The origin of the coordinate system is at the center of the sphere that corresponds to the earth's *center of mass*. The x-axis lies in the equatorial plane with its positive end intersecting the Greenwich meridian. The y-axis lies in the equatorial plane with its positive end intersecting the sphere at 90° E longitude. The z-axis is coincident with the earth's spin axis, positive toward the North Pole.

Coordinate Transformations

Coordinate transformations are required to convert from curvilinear coordinates to Cartesian coordinates and vice-versa. The transformations are presented below:

Direct Transformation
{x, y, z} = f {ψ, λ, h_s}

Inverse Transformation
{ψ, λ, h_s} = g{x, y, z} where g = f^{-1}

$x = (R + h_s) \cos\psi \cos\lambda$

$\lambda = \arctan (y / x)$

$y = (R + h_s) \cos\psi \sin\lambda$

$\psi = \arctan [z / (x^2 + y^2)^{1/2}]$

$z = (R + h_s) \sin\psi$

$h_s = (x^2 + y^2 + z^2)^{1/2} - R$

Azimuths and Distances on the Sphere

It is useful to be able to compute the azimuth and distance between two points on the sphere. An azimuth on the sphere is the spherical angle between any great circle and a meridian. For instance, figure 3.5 shows points A and B, both located on the sphere $(h_s = 0)$. How does one compute the azimuth from A to B? From B to A? Do they differ by exactly 180°? How far is it from A to B?

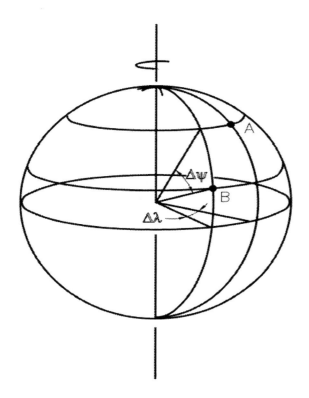

Figure 3.5 Azimuth and Distance from A to B.

Is the distance measured along a great circle? Small circle? Chord? The answer to these questions can be found by constructing a spherical triangle that includes A, B and a pole. Before we address this problem, it would be wise to review spherical triangles.

Spherical Triangles

A spherical triangle is a triangle located on the surface of a sphere, the sides of which are formed by great circle arcs. Unlike plane triangles, the sides and included angles of a spherical triangle are expressed using arc measure. That means the sides of the triangle are expressed as the included angle (measured at the center of the sphere) that subtends the side. The relationship $(s = r\,\theta)$ will come in handy when dealing with spherical triangles. Figure 3.6 shows a generic spherical triangle having included angles A, B and C and sides a, b and c opposite their corresponding included angle.

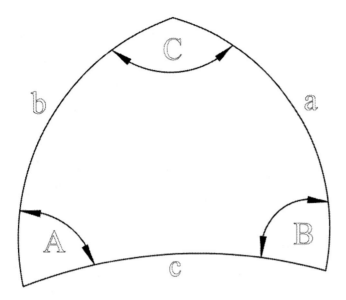

Figure 3.6 Typical Spherical Triangle.

Spherical Trigonometry

The relationships between angles and sides in spherical triangles are governed by *spherical trigonometry*, the laws of which are similar in form to plane trigonometry. Some of the more useful relationships are presented below.

Law of Sines

$$\frac{\sin a}{\sin A} = \frac{\sin b}{\sin B} = \frac{\sin c}{\sin C}$$

It should be noted that the law of sines will be ambiguous for angles in excess of 90° due to the fact that the sine function is positive in both the 1st and 2nd quadrants. Therefore, the law of sines is frequently not used in favor of the law of cosines.

Law of Cosines

There are two forms for the law of cosines - one for sides and one for angles. Note that the forms are both *cyclic*, i.e., a family of formulas may be obtained by cycling the side and angle variables.

$\cos a = \cos b \cos c + \sin b \sin c \cos A$ 　　(law of cosines for sides).

$\cos A = -\cos B \cos C + \sin B \sin C \cos a$ 　(law of cosines for angles)

Problem 3.2: Calculate the great circle distance between Points A and B in Problem 3.1. The great circle trace between Point A and Point B is different than the arc traced by the parallel of latitude. Refer to figure 3.7. Since the sides of a spherical triangle are great circles, the solution of a spherical triangle can give the great circle distance between Points A and B. Figure 3.8 shows detail of the spherical triangle used to solve this problem.

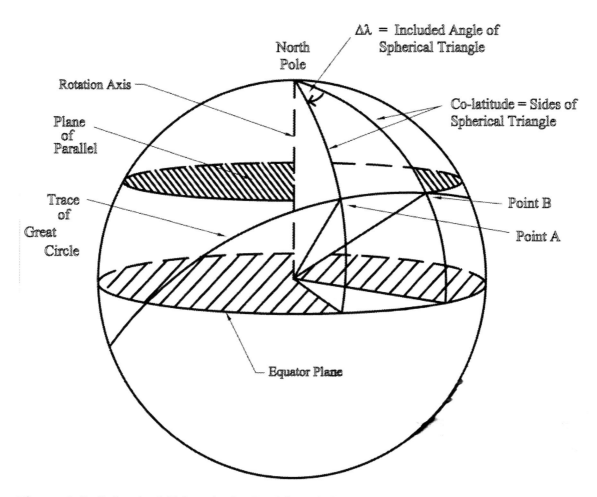

Figure 3.7 Spherical Triangle for Problem 3.2.

Analysis of this problem finds data for two sides and the included angle. The value of the included angle is $\Delta\lambda$ = 2° 10' 40" , computed in Problem 3.1. The value of the two known sides equals the *co-latitude* (δ, delta) of Point A and B. Co-latitude = 90° - latitude. Figure 3.9 illustrates the co-latitude. The co-latitude for both Point A and B is 90° - 42° 15' 16" = 47° 44' 44". The distance between Point A and Point B on the trace of the great circle is a side of the

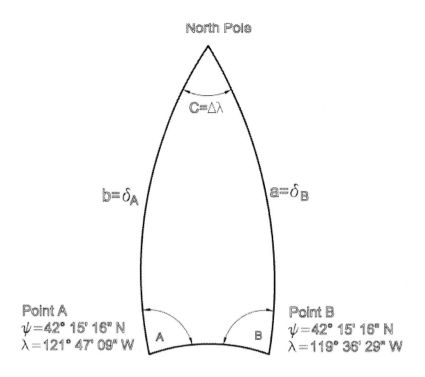

North Pole

$C=\Delta\lambda$

$b=\delta_A$ $a=\delta_B$

Point A
$\psi=42°\ 15'\ 16"\ N$
$\lambda=121°\ 47'\ 09"\ W$

A   B

Point B
$\psi=42°\ 15'\ 16"\ N$
$\lambda=119°\ 36'\ 29"\ W$

Figure 3.8 Spherical Triangle Detail.

spherical triangle as shown in figure 3.7. We can use the Law of Cosines: $\cos c$ = $\cos a \cos b + \sin a \sin b \cos C$. The computations follow figure 3.9.

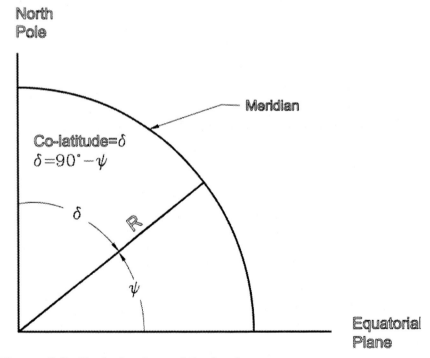

North Pole

Meridian

Co-latitude=δ
$\delta=90°-\psi$

δ

R

ψ

Equatorial Plane

Figure 3.9 Co-latitude and Latitude.

cos (distance AB) = cos (47° 44' 44") cos (47° 44' 44") + sin (47° 44'44") sin (47° 44'44") cos (2° 10' 40") = 0.999604307512.

Angular Distance = arcos(0.999604307512) = 1°36' 42.74"

Angular Distance = 1°36' 42.74" ⇒ 1.61187313° (π/180°) = 0.0281324932 radians

The Law of Cosines gives the distance between Point A and B as an angular distance. To compare our answer to that of Problem 3.1, let us convert this angular distance to a lineal distance. We have to remember that the radius of a great circle is the same as the sphere itself.

s = R (Angular Distance) = 6,371,000 (0.0281324932) = 179,232 meters

So is a great circle trace between two points shorter than the distance on the parallel of latitude? In Problem 3.1 we computed a distance on the parallel of 179,237 meters. The distance between the same two points on the great circle is 179,232. This is a difference of 5 meters.

Cotangent Formulas

Two cotangent formulas, which may be derived from the laws of sines and cosines, are extremely useful in our applications:

$$\cot A = \frac{\cot a \sin b - \cos b \cos C}{\sin C} \qquad \cot B = \frac{\sin a \cot b - \cos a \cos C}{\sin C}$$

The reader is referred to Mackie (1985) for a derivation of these formulas.

Normal Sections

A *normal section* is a plane that contains the normal to the sphere at the occupied point and another point of interest (backsight, foresight, pole, etc.). A horizontal angle, measured with a theodolite or total station, is the angle between two normal sections at the instrument location. An azimuth is measured from the normal section containing the North Pole clockwise to the normal section containing another point. All normal sections on the sphere intersect the sphere along great circle arcs (by definition).

Problem 3.3: Calculate the azimuths between points A and B using the data from Problems 3.1 and 3.2.

Since an azimuth is defined as the clockwise angle measured from North, the solution of the included angle A at Point A (refer to figure 3.8) is the azimuth from point A to point B. The cotangent formula is handy for this solution:

$$\cot A = \frac{\cot a \sin b - \cos b \cos C}{\sin C}$$ or, you may wish to write it like this:

$$\tan A = \frac{\sin C}{\dfrac{\sin b}{\tan a} - \cos b \cos C}$$

$$\tan A = \frac{\sin 2°10'40''}{\dfrac{\sin 47°44'44''}{\tan 47°44'44''} - \cos 47°44'44'' \cos 2°10'40''}$$

$$\tan A = \frac{0.038000}{0.000486} = 78.242628$$

By inspection of algebraic signs of numerator and denominator, angle A is located in the first quadrant; therefore,

Angle A = arctan 78.242628 = 89.267756° = 89°16'03.9" = Azimuth point A to point B

Since this particular spherical triangle is isosceles (sides a and b are equal), the included angle *B* at point B also equals 89°16'03.9" and the azimuth from point B to point A = 360° - 89°16'03.9" = 270°43'56.1". Notice that the difference in azimuths is not exactly 180°. Why is this? (Hint: Keep reading!)

Other Spherical Earth Characteristics

When the surveyor considers the fact that the earth is not a plane, but rather a curved surface, he or she must be aware of the characteristics associated with a curved earth.

Convergence of the Meridians

All meridians converge at the poles. Since all meridians are north-south lines, it is by necessity that the bearing of a line as measured from one end does not equal the bearing ± 180° as measured from the other end of the line. The effect of meridional convergence may be significant, particularly on long east-west traverses. The magnitude of this convergence (θ) may be estimated by:

$$\theta'' = \frac{\rho \, \overline{d} \, \tan \overline{\psi}}{R}$$

where θ" is the convergence of the meridians expressed in arc seconds, $\rho = (180 * 3600)/\pi$ seconds per radian, $\overline{d}$ is the east-west distance between the two points, as measured along a parallel circle at the mean latitude, and $\overline{\psi}$ is the mean latitude on the sphere. Note the Greek lowercase letter (θ) theta is pronounced (THAY-tuh) and the Greek lowercase letter (ρ) rho is pronounced (row).

Spherical Excess

The summation of the included angles in a spherical triangle will always exceed 180° by an amount known as the *spherical excess*. Accordingly, the summation of the interior angles in a spherical triangle = A + B + C = 180° + ε where ε is the spherical excess. Spherical excess is proportional to the area of the spherical triangle. Several formulas provide the amount of spherical excess. One formula for spherical excess is in terms of all three sides.

$$\tan^2 \frac{1}{4}\varepsilon = \tan\frac{1}{2}s\left(\tan\frac{1}{2}(s-a)\right)\tan\frac{1}{2}(s-b)\left(\tan\frac{1}{2}(s-c)\right) \text{ where } s = \frac{1}{2}(a+b+c)$$

An estimate for spherical excess is given by:

39

$$\varepsilon = \frac{bc \sin A}{2R^2 \sin 1''}$$

(*b* and *c* are expressed in metric units, not arc units) What are the units for spherical excess as stated in this equation?

Problem 3.4: Consider the spherical triangle formed in Problem 3.2. Compute the value for the spherical excess of the spherical triangle.

Solution #1: First, make use of the equation for interior angles in a spherical triangle. We have already computed angles *A* and *B*. Note that these angles are not equal to 90°. While the parallel of latitude meets the two meridians at right angles the meridians do not meet the great circle trace at right angles.

A + B + C = 180˚ + ε ⇒ 89° 16' 03.92" + 89° 16' 03.92" + 2° 10' 40" + ε

ε = 0° 42' 48"

Solution #2: An example computation using the formula in terms of the three sides for spherical excess follows. To use the formula we need to compute (*s*):

$$s = \frac{1}{2}(a+b+c) =$$

$$s = \frac{1}{2}\left(1°36'42.74''+47°44'44''+47°44'44''\right) = 48°33'05.37''$$

$$\tan\frac{1}{2}s = 0.451007795032$$

$$\tan\frac{1}{2}(s-a) = 0.434165978271$$

$$\tan\frac{1}{2}(s-b) = \tan\frac{1}{2}(s-c) = 0.00703323531598$$

$$\tan^2\frac{1}{4}\varepsilon = 0.00000968612642146$$

$$\tan\frac{1}{4}\varepsilon = 0.00311225423471$$

$$\frac{1}{4}\varepsilon = 0.178318456684$$

ε = 0.7132738° or 0° 42' 48"

Rule of Thumb

A useful rule of thumb is that a one arc second change in latitude or longitude corresponds with a ground distance of approximately 30 meters. Can you prove this?

Study Questions

Note: For the following problems use a reference sphere with a radius of 6, 371, 000 meters.

1. In March 2000, the Annual Convention of the American Congress on Surveying and Mapping was held in Little Rock, Arkansas (ψ = 34° 43' 21" N; λ = 92° 21' 15" W).

   a) Compute the distance in kilometers from Troy, Alabama (ψ = 31° 48' 02" N; λ = 85° 57' 26" W) on the sphere.
   b) Compute the spherical excess of the spherical triangle in a) using two different formulas. Compare your answer.
   c) Compute the distance to the Convention in miles from Klamath Falls, Oregon (: ψ = 42° 15' 16"N; λ = 121° 47' 09"W)
   d) Compute the spherical excess of the spherical triangle in c) using two different formulas. Compare your answer.

2. Compute the lineal distance (in meters) covered by one second of latitude on the sphere.
3. At what latitude will a second of latitude equal a second of longitude? Prove your answer.
4. Write the cyclic equation for Angle C in Problem 4. What does cyclic mean?
5. Consider a point P on the sphere at a latitude of 31°48' 02" and a west longitude of 85° 57' 26". Compute the length in meters of a second of latitude and a second of longitude at point P.
6. A student at Troy State University is considering going to visit the famous geyser in Yellowstone National Park referred to as "old faithful." The latitude and longitude of Troy State University is given in question one. The latitude and longitude of the geyser is ψ = 44° 27' 38" and λ = 110° 49' 41" W. Compute the shortest distance on the sphere from Troy State University to the geyser.
7. Let us assume that the elevation above the sphere (h_S) at Troy State University is 150.000 meters and the elevation above the sphere at the famous geyser is 2805.000 meters. Using the coordinate transformation from curvilinear to rectangular coordinates, compute the rectangular coordinates for Troy State University and the geyser. Compute the 3D vector (Δx, Δy, Δz) and the mark-to-mark (chord) distance between the two locations. Explain the difference between the distance computed on the sphere and the length of the vector between TSU and the geyser.

8. Compute the difference between the lineal distance of the arc which subtends a second of **longitude** at ψ = 31° 48' 02"N and a second of **longitude** at ψ = 42° 15' 16". Justify your answer.

9. Compute the difference between the lineal distance of the arc which subtends a second of **latitude** at ψ = 31° 48' 02"N and a second of **latitude** at ψ = 42° 15' 16". Justify your answer.

10. Compute the azimuth from Troy State University to the geyser discussed in question number 6. Compute the back azimuth. Is the difference between the azimuth and back azimuth a function of 180°. Why or why not?

11. Describe a spherical triangle where the summation of the interior angles = 270°.

CHAPTER FOUR

THE EARTH'S GRAVITY FIELD

The gravity field of the earth consists of two parts, the principal one caused by the attraction according to Newton's law, the second one caused by the earth's rotation. At the equator the second part is about 1/3 per cent of the first; elsewhere it is less.
Heiskanen and Meinesz, 1958.

Our continued investigation into geometric geodesy will detour with a look into physical geodesy, in particular -- gravity. This will set the stage for further discussions of geometric geodesy, using an ellipsoid as the geometric reference surface. We will return to the topic of physical geodesy later as we express some relationships between geometric and physical geodesy.

Gravity

What is gravity? A force? An acceleration? Why is gravity important to surveyors? Isn't it enough to know "what goes up must come down"? Gravity defines an important reference direction for surveyors -- the *plumb line* (local vertical) is defined by gravity. Gravity also defines an important reference surface -- a *level surface* is a surface that is perpendicular to the plumb line at all points. What is the difference between a level surface and a *horizontal plane*?

Gravitation

To define gravity we must first start with *gravitation*. You may remember this term from your studies in physics. We will go a bit deeper here in order to provide us with a meaningful expression for gravitation. Let's begin with an expression for force: $F = ma$; where F equals force, m equals mass and a equals acceleration. This simple expression gives us a useful understanding about forces—they are the result of masses acted upon by accelerations.

Newton developed his *law of gravitation* in 1687. It describes the attracting force present between two masses:

$$F = \frac{Gm_1m_2}{l^2}$$

F is the attracting (gravitational) force between point masses m_1 and m_2 located at a distance, l, apart. The *gravitational constant, G,* is one of the least accurate physical constants, having a relative accuracy of approximately 1×10^{-4}. The currently accepted value for G is 6.67259×10^{-11} m³ kg^{-1} s^{-2}.

Now suppose we replace one of the point masses in Newton's equation with the mass of the earth, M. Dropping the subscript for the remaining point mass, m, we can rewrite the equation as:

$$F = \frac{GMm}{l^2}$$

The accepted value of the constant GM is $398,600.5 \times 10^9$ m³ s^{-2} per the Geodetic Reference System of 1980 (GRS 80) adopted by the International Union for Geodesy and Geophysics (IUGG). The equation assumes that the earth's mass is located at a finite point—the *center of mass.*

Now let's combine this equation with our first equation and solve for acceleration:

$$a = \frac{GM}{l^2} = b$$

The term "b" is the gravitational acceleration caused by the earth's mass or, simply, *gravitation.* Note that this is a scalar expression, it yields magnitude only. So, gravitation is an acceleration that will cause a force on a mass located within the earth's gravitational field: $F = mb$.

Problem 4.1: Determine the magnitude of the gravitational acceleration on the surface of the earth using a spherical earth geometric reference surface.

$$b = \frac{398,600.5 \times 10^9 \text{ m}^3/\text{s}^2}{(6,371,000 \text{ m})^2} = 9.82 \text{ m/s}^2$$

A mass located within the earth's gravitational field is going to be accelerated toward the earth's center of mass. This characteristic will allow us to express gravitation as a vector quantity possessing magnitude and direction. We will express the vector form of gravitation as $\vec{b}$ which equates to :

$$\vec{b} = \frac{GM}{l^2} \frac{\vec{l}}{l}$$

The term "$\vec{l}$" represents the position vector of the point mass with respect to the earth's center of mass and $l = |\vec{l}|$.

Centrifugal Acceleration

Remember your playground days? Do you remember the sensation of being spun around on the merry-go-round? *Centrifugal acceleration* threatened to throw you off the apparatus if you didn't hold on tight. Remember how you had to hold on tighter the farther you got from the center of the merry-go-round and how you didn't have to hold on at all when at the center? A spinning earth creates the same centrifugal acceleration.

Like gravitational acceleration, centrifugal acceleration is a vector quantity. The direction of the acceleration is always perpendicularly outward from the rotation axis. The formula for centrifugal acceleration, $\vec{z}$, is $\vec{z} = \omega^2 \vec{p}$ where ω is the angular velocity in radians per time unit and $\vec{p}$ is the vector from spin axis.

Problem 4.2: Determine the magnitude of the earth's centrifugal acceleration at the equator and at the north pole. Assume a spherical earth model as in Problem 1.

First, we must determine the earth's angular velocity in radians per second.

ω = 360°/24 hr (π radians/180°) (1 hr/3600sec) = 0.000073 radians/sec

At the equator, the magnitude of the vector from the spin axis to the surface of the geometric reference surface equals the radius, therefore:

$\vec{z}_{equator}$ = (0.000073 radians/sec)² (6,371,000 m) = 0.03 m/s²

At the pole (north or south), the magnitude of the vector from the spin axis is zero since the north pole is defined as the intersection of the spin axis with the sphere surface. Accordingly, $\vec{z}_{pole} = 0$ m/s²

Gravity Acceleration

Gravity acceleration, $\vec{g}$, or gravity is the vector sum of gravitational acceleration and centrifugal acceleration: $\vec{g} = \vec{b} + \vec{z}$. Refer to figure 4.1. The direction of the plumb line is defined by $\vec{g}$. Notice that the plumb line will not be directed toward the earth's center of mass, except at the poles. Which is greater, $\vec{g}_{equator}$ or $\vec{g}_{pole}$? Why? Could changes in gravity affect the performance of Olympic athletes? Gravity is measured in *gals* where 1 gal = 1 cm/s². Therefore, 981 gal = 9.81 m/s². This model of gravity is somewhat simplistic because gravity is also affected by tidal accelerations, earth tides and local density variations.

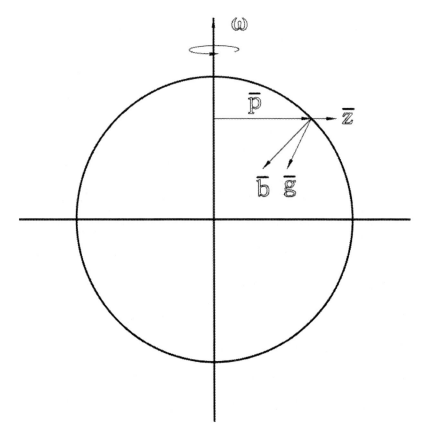

Figure 4.1 Gravitational and centrifugal acceleration vectors.

Potential

Potential is a function of position in a force field, given by the line integral over a path from one reference point to another under the condition that the potential is path independent. For most force fields, the potential is zero at infinity. Potential is a scalar quantity (magnitude only]. Refer to figure 4.2. Potential at point P is defined by the formula:

$$\int_{\infty}^{p} \vec{f} d\vec{l}$$

$\vec{f}$ is the force dependent upon position within the force field. $d\vec{l}$ is an element of distance. Note that the potential is expressed in units of work and that it is path independent.

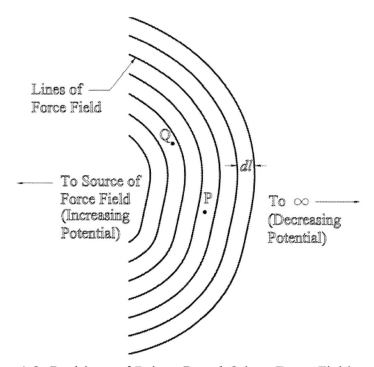

Figure 4.2 Positions of Points P and Q in a Force Field.

The change in potential (work required to move) from point P to point Q within a force field is:

47

$$\int\limits_{P}^{Q} \vec{f} d\vec{l}$$

Gravitational Potential

Gravitational potential, *V*, is the potential, attributable to the presence of a gravitational field, caused by an attracting mass. The attracting mass may be some mass *m* or the mass of the earth *M*. The magnitude of this potential indicates the work that must be done by the gravitation in order to move a unit mass from infinity (*V* = 0) to the point under consideration (Torge, 1991). The gravitational potential caused by point mass *m* may be formulated as:

$$V = \frac{Gm}{l} \quad \text{with } \lim_{l \to \infty} V = 0.$$

Since *l* is the magnitude of the position vector from *m*, then,

$$l = \sqrt{x^2 + y^2 + z^2} \text{ and } V = \frac{Gm}{\sqrt{x^2 + y^2 + z^2}}.$$

The earth is composed of an infinite number of mass elements, *dm*. Theoretically the gravitational potential of the earth (taken from the potential of a solid body) is

$$V = G \iiint \frac{dm}{l}.$$

The equation shows that we must account for each and every mass element comprising the earth in order to determine the earth's gravitational potential at any distance, *l*, from the center of mass. Not a simple task considering the density variations within the earth. If we consider the earth to be spherical and composed of homogeneous layers, the potential at a point not within the earth may be expressed as

$$V = \frac{GM}{l}$$

which considers the mass of the earth to be concentrated at a point, the center of mass. Note that gravitational potential is dependent on position within the gravitational force field, *V* = *f*(*l*). Consider an object in the earth's gravitational field: the object has potential, *V*, by virtue of its position within the force field, it experiences an acceleration, $\vec{b}$, due to the force field (irregardless of its mass), and the object's mass causes it to be attracted to the earth with force *F*.

Centrifugal Potential

The *centrifugal* or *rotational potential*, *Φ*, of a point due to its position with respect to the rotational (spin) axis is given by: $Φ = ½ ω^2 p^2$ where *ω* is the angular velocity and *p* is the distance from the rotational axis. This may be derived by integrating centrifugal force, $F_ω$, as follows:

$F_ω = ma_ω = mz = mω^2 p$

Let *m* be a unit mass (m = 1),

$$\therefore F_ω = ω^2 p \text{ where } p = \sqrt{x^2 + y^2}$$

$$Φ = \int ω^2 p\, dp = \frac{1}{2}ω^2 p^2 = \frac{1}{2}ω^2(x^2 + y^2)$$

Gravity Potential

Gravity potential, W, is the sum of the gravitational and centrifugal potentials:

$$W = V + Φ = \frac{Gm}{\sqrt{x^2 + y^2 + z^2}} + \frac{1}{2}ω^2(x^2 + y^2).$$

Equipotential Surfaces

An *equipotential surface* is defined as a surface having constant gravity potential, **not** constant gravity. Equipotential surfaces are also known as *level surfaces* or *geopotential surfaces*. These surfaces are perpendicular at all points to the gravity vector (plumb line). There are an infinite number of equipotential surfaces that may be considered. A still lake surface is an equipotential surface. This lake surface is **not** horizontal, level surfaces are curved. Equipotential surfaces have the following properties (Vanicek & Krakiwsky, 1982):

1. They are closed, continuous surfaces that never cross one another.
2. They are formed by long radius arcs, in general, without abrupt steps.

3. They are convex everywhere, they do not possess troughs, though it is nearly impossible to illustrate them as such.

The Geoid

The *geoid* is the fundamental equipotential surface of physical geodesy. The geoid may be defined as: *the equipotential surface of the earth's gravity field which best fits, in a least squares sense, mean sea level.* The geoid would coincide with the ocean surface if the oceans were affected only by gravity and not disturbed by tides, currents, winds, atmospheric pressure, etc. Figure 4.3 shows the relationship between the geoid and equipotential surfaces. Note that the equipotential surfaces are not, in general, parallel. Since the plumb line must be perpendicular to a level surface, this results in curvature of the plumb lines.

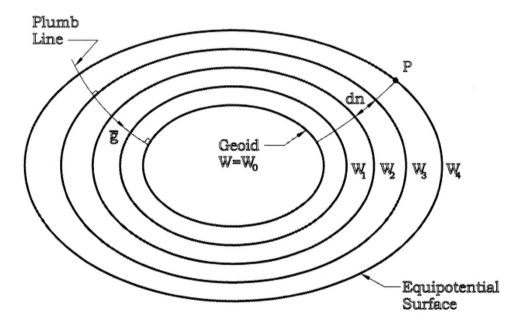

Figure 4.3 Equipotential Surfaces.

Gravity

We have previously defined gravity in terms of gravitation and centrifugal acceleration. Gravity may also be defined as the radial rate of change of gravity potential. The magnitude of gravity (g) may be determined from figure 4.3 by the following expression:

$$g = \left|\vec{g}\right| = \frac{-dW}{dn}$$

where dW is the incremental change in gravity potential and dn is the incremental change in vertical distance (measured along the plumb line).

Geopotential Number

The *geopotential number, C* (also *GPN*), is a value derived from the difference in gravity potential between an equipotential surface under consideration and the geoid:

$$C = W_0 - W_p = \int_0^P g\, dn$$

Note that the geopotential number represents the work required to move a 1 kg mass from the geoid to the equipotential surface at P. The geopotential number at a point is approximately (2% smaller) equal numerically to the *elevation* of that point in meters. The reason for the discrepancy lies in the gravity value used: geopotential numbers are based upon actual (measured) gravity while elevations (NGVD 29) are based upon *normal gravity* that is a latitude-dependent quantity and an estimate of actual gravity.

Orthometric Height

The *orthometric height* of a point is the vertical distance from the geoid to that point. Once again note that vertical distance is measured along the (curved) plumb line. Orthometric height and elevation will often be considered synonymous for our discussions. In reality there is a difference. The orthometric height (H) may also be determined from:

$$H = \frac{C}{\bar{g}}$$

where C is the geopotential number of the point and $\bar{g}$ is the mean gravity along the plumb line. Since mean gravity cannot be measured directly, a hypothesis is made regarding the mass distribution within the earth in order to estimate this value.

Dynamic Height

Another height definition is the *dynamic height*. This height is computed using the geopotential number in a manner similar to that shown above for orthometric heights except that a value for sea level normal gravity at latitude 45° is used in place of mean gravity. While not used by surveyors, you may encounter this height on data sheets for vertical control stations.

Does Water Always Flow Downhill?

We take it for granted that water flows downhill, but is it possible that water could actually flow *uphill*? There are places on earth where natural oddities such as this do occur. Fortunately, such an oddity can be explained by our knowledge of equipotential surfaces. Figure 4.4 shows the typical occurrence of a ball rolling downhill. Why does the ball roll downhill (from A to B)? Is it because of the topography (A is "higher" than B) or gravity potential?

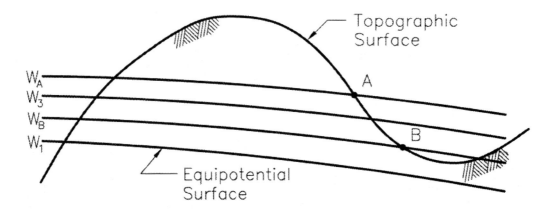

Figure 4.4 Downhill explained in terms of potential instead of topography.

Since A possesses a lesser gravity potential than B (as evidenced by the equipotential surfaces), the ball will roll from A to B. (Note: this may seem to be the opposite of potential energy examples given in physics class, but remember $\lim_{l \to \infty} V = 0$.)

Now consider figure 4.5. Assume the orthometric height of D is greater than that of C. But the figure shows that C has a lesser gravity potential than D. The ball must roll from a point of lower potential to a point of higher potential; therefore, the ball will roll "uphill" from C to D! Gravity potential determines relative highs and lows, not geometry (orthometric heights).

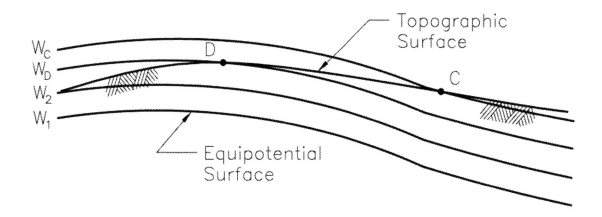

Figure 4.5 Example where a ball will roll uphill against topography.

Leveling

Leveling refers to *differential leveling, spirit leveling,* or *geometric leveling.* Leveling is performed by surveyors in order to determine orthometric height differences between two or more points. Orthometric height differences are provided by leveling only where there is parallelism between the equipotential surfaces. In general, such parallelism cannot be assumed and leveling provides geometric height differences, not orthometric height differences. Keep in mind that this will typically be a problem only for surveyors where they are performing precise leveling extending over a significant distance.

Single Setup

Figure 4.6 represents the scenario for a single instrument setup used to determine the orthometric height difference between points A and B. Though the equipotential surfaces are not parallel, since we are covering such a small distance on the earth's surface (AB is a "quasi-differential" length) we can assume that they are parallel. Accordingly, we can state the following:

$$\delta n_{AB} = BS - FS = H_B - H_A = \Delta H_{AB}$$

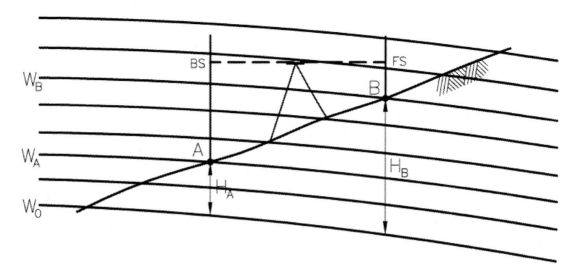

Figure 4.6 Single Setup: Assumption of parallelism between equipotential surfaces.

where δn_{AB} is the leveled (incremental) height difference between A and B that is equal to the orthometric height difference between A and B (ΔH_{AB}).

Multiple Setups - Leveling Circuit

Suppose we desire to determine the orthometric height difference between points A and Z as shown in figure 4.7. If we are performing precise leveling, we can no longer neglect the fact that the equipotential surfaces are not parallel. We have defined the orthometric height of a point in two ways: geometrically as the vertical distance from the geoid; and physically as a relationship between the geopotential number and mean gravity. Since it is impractical to measure the vertical distance from the geoid (how does one locate the geoid?!), we will focus on the physical definition. Accordingly, the orthometric height difference between A and Z may be expressed as:

$$\Delta H_{AZ} = H_Z - H_A = \left[\frac{C_Z}{\overline{g}_Z} - \frac{C_A}{\overline{g}_A} \right] = \left[\left[\frac{W_0 - W_Z}{\overline{g}_Z} \right] - \left[\frac{W_0 - W_A}{\overline{g}_A} \right] \right]$$

Note: $\overline{g}_z \neq \overline{g}_A$

54

This expression tells us we need either the gravity potentials at A and Z or the difference in gravity potential between the two points ($\Delta W_{AZ} = W_Z - W_A$). We can relate our previous expression for the magnitude of gravity and our single setup leveled height difference by the following:

$$g = |\vec{g}| = \frac{-dW}{dn} = \frac{-\delta W}{\delta n} \cong \frac{-\delta W}{\delta H}$$

where δn is the single setup leveled height difference and δW is the increment of gravity potential between the two-leveled points. Rearranging and substituting we have:

$$-\delta W = g \delta n \cong g\, \delta H \cong \overline{g}_Z \delta H_Z$$

where $\overline{g}_Z$ is the magnitude of gravity at Z and δH_Z is the incremental orthometric height at Z. It should be noted that:

$$\delta H_Z = \frac{g}{\overline{g}_Z} \delta n \neq \delta n \qquad \text{therefore:} \quad \delta n \neq \delta H_Z \neq \delta H_A$$

Which proves that orthometric height differences (δH_Z, δH_A) cannot be obtained by leveled height differences (δn) alone; i.e.:

$$\Delta H_{AZ} = H_Z - H_A \neq \sum_{A}^{Z} \delta n$$

As stated previously, differences in gravity potential are required in order to determine true orthometric height differences: $\delta W = -g\delta n$

$$\Delta W_{AZ} = W_Z - W_A = -\sum_{A}^{Z} g\delta n = -\int_{A}^{Z} g\, dn \text{ (must make gravity measurements!)}$$

55

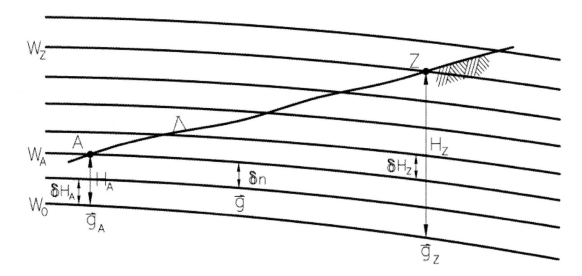

Figure 4.7 Multiple Setups: non-parallelism of equipotential surfaces cannot be ignored.

Level Loops

When performing precise leveling over a closed loop, the change in gravity potential will be zero but the summation of leveled height differences will not be zero! ($\oint$ = integral over a closed loop).

$$\oint g\,dn = \oint dW = 0$$

$$\oint \delta n \neq 0 = \varepsilon \quad (\varepsilon = \text{orthometric excess})$$

Orthometric Excess & Correction

Orthometric excess occurs because the equipotential surfaces are not parallel. In general, the equipotential surfaces are not parallel in a north - south direction but they are parallel in an east - west direction. This generality does not, however, account for local gravity variations. An *orthometric correction* may be applied to convert leveled height differences into orthometric height differences. The orthometric correction is only a general correction and cannot account for local gravity variations. The orthometric correction may be estimated using the following formula from NGS Manual 3:

$$O.C. = (-2\alpha)\overline{H}\sin(2\overline{\phi}) * \{1 + [\alpha - (2\beta/\alpha)]\cos(2\overline{\phi})\} * \Delta\phi$$

56

where *O.C.* is the orthometric correction, $\alpha = 0.002644$, $\beta = 0.000007$, $\overline{H}$ is the mean orthometric height of the leveling section , $\overline{\phi}$ is the mean latitude of the section, and $\Delta\phi$ is the change in latitude for the section (expressed in radians). The orthometric correction should be computed on sections of 5 - 10 km in length. Where the leveling circuit exceeds that length, it should be broken into sections.

Anderson and Mikhail, 1998 provide a second formula for computing the orthometric correction in terms of the orthometric elevation and latitude of the starting point.

$$O.C. = (-0.005288)\sin 2\phi\, H\, \Delta\phi\,(arc\,1')$$

Where ϕ is the latitude of the starting point, H is the orthometric height at the starting point, and $\Delta\phi$ is the change in latitude in minutes between the two points. $\Delta\phi$ is positive in the direction of increasing latitude. Arc 1' is one minute expressed in radian measure.

Problem 4.3: A differential level loop is surveyed around a 10 km square, beginning at point A as shown in figure 4.8. Determine the orthometric correction for each side of the level loop.

Line AB: By inspection, the change in latitude is zero; $O.C._{AB}$ = 0 m.
Line BC: Using the second formula:

$$O.C._{BC} = (-0.005288)\sin[2(42°00'00'')](500\,\text{m})(5.4')\left(\frac{1'}{60'/°}\right)\left(\frac{\pi}{180°}\right) = -0.0041\,\text{m}$$

Line CD: By inspection, the change in latitude is zero; $O.C._{CD}$ = 0 m.
Line DA: By inspection, the orthometric height of the starting point is 0 m; therefore, $O.C._{DA}$ = 0 m.

The orthometric correction for the entire level loop is –0.0041 m.

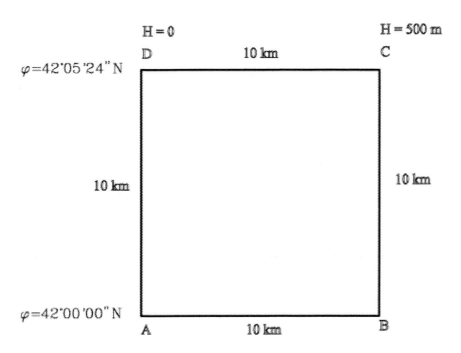

Figure 4.8 Level loop for orthometric correction example.

Study Questions

1. Sir Isaac Newton was able to predict that the earth was flattened at the poles without leaving his house to obtain measurements. This prediction was subsequently supported by extensive field measurements in Peru and in Lapland. Upon what reasoning was his analysis based?

2. A lake is one hundred miles long in its north south extent. On a calm day a surveyor set a level between the southern-most edge of the lake and a nearby benchmark that has an NAVD88 elevation of 302.36 feet. The backsight reading on the benchmark was 1.56 feet. The foresight reading on the rod set at the shoreline at waters edge was 8.52 feet. The scaled latitude at the benchmark is 34° 19'. Calculate the expected elevation of the lake at its northernmost edge by assuming that every second of latitude you travel averages 30 meters on the ground.

3. When a surveyor levels up an instrument such as a total station, to what is the level vial referenced?

4. Why is it possible to find situations where a ball will roll uphill with respect to topography?

5. If you were expected to set the next world's record for the high jump and you could pick the place for the next Olympic trials, what would you choose:
    a) A place at a high elevation or a low elevation?
    b) A place near the equator or nearer the North Pole?

6. A crew ran a precise level loop beginning at a benchmark on the bottom of a USGS 7 1/2 minute map at a latitude of 30° 30' to a point at the top of the map. The elevation of the benchmark is 120.751 meters. The difference in

elevation between the benchmark and the point was computed to be 1420.95 feet. Calculate the orthometric correction necessary before the spirit-leveled elevation at the top of the map can be used as a benchmark.

7. Consider the fact that equipotential surfaces are not parallel. What impact does this fact have on spirit leveling?

8. Must points of equal orthometric height be located on the same equipotential surface? Defend your answer with a discussion.

9. Assume that an extremely dense mass of finite size is located just beneath the ocean floor. If the mass is substantially denser than the surrounding ocean, will the geoid (ocean surface) be raised or depressed over the mass?

10. In general, are equipotential surfaces parallel north-south or east-west? Why?

CHAPTER FIVE

GEOMETRY OF THE ELLIPSOID

If the earth were only slightly flattened, a sphere would be an ideal mathematical surface. It is a simple figure defined by only one parameter, its radius. As it is, the earth's departure from a true sphere is too great and so the thought of a sphere must be discarded.

Some other mathematical figure must be adopted. It must be simple enough so the computations are not overly difficult, but must nowhere depart from the true figure of the earth by an amount which will give intolerable errors in the results. This figure is the ellipsoid of revolution, generally referred to simply as the ellipsoid; it is produced by rotating an ellipse about its minor axis, with the major axis generating the equatorial plane. This ellipsoid of revolution approximates an oblate spheroid and the terms "ellipsoid of revolution" and "spheroid" are used interchangeably.
      Ewing and Mitchell, 1970

The use of an *ellipsoid of revolution* as a mathematical model for the earth provides a more accurate representation than that afforded by the spherical earth model. The ellipsoid of revolution is formed by rotating a *meridian ellipse* about its minor axis, thereby forming a three-dimensional solid, the ellipsoid. The ellipsoid model chosen should closely agree with the geoid as determined by a *least squares best fit*. The terms "ellipsoid of revolution" and "ellipsoid" will be used interchangeably in the chapters that follow.

Meridian Ellipse

Figure 5.1 shows a typical ellipse. This particular ellipse is a 'meridian ellipse' formed by the intersection of a normal section (plane) containing the earth's rotation axis and the ellipsoid. The parameters of an ellipse are:

a: length of semi-major axis

b: length of semi-minor axis

f: flattening = $(a - b) / a$

e: first eccentricity = $[(a^2 - b^2) / a^2]^{1/2} = (2f - f^2)^{1/2}$

e':   second eccentricity = $[(a^2 - b^2) / b^2]^{1/2}$

An ellipse may be defined by any two parameters, however the convention in geodesy is to define the meridian ellipse, hence the ellipsoid, by a and f.

The ellipse has two focus points, or *foci*, located along the major axis. The distance from the *origin* (intersection of major and minor axes) to a focus point is *ae*. The sum of the distances from each focus point to any point on the ellipse is constant. Knowing this fact, an ellipse may be readily constructed by holding two ends of a string in place (at the foci) and using a pencil stretch the string tight while drawing the ellipse.

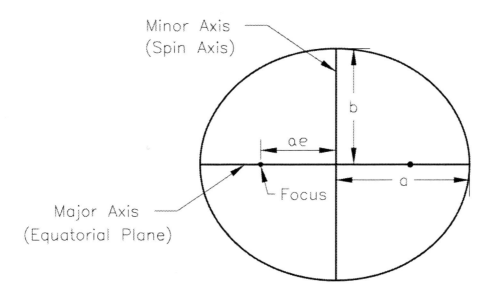

Figure 5.1 Two-dimensional ellipse in Meridian Section.

By inspection one can see that the distances from each focus point to the intersection of the minor axis with the ellipse are equal. This means that the distance from each focus to the intersection of the minor axis with the ellipse equals the distance (a) or the length of the semi-major axis. Since the minor axis is perpendicular to the equatorial axis, a right triangle can be extracted for analysis. Consider the right triangle consisting of the three sides with lengths $ae, b,$ and a shown in figure 5.2.

Side $ae = \sqrt{a^2 - b^2}$ or $e = \dfrac{\sqrt{a^2 - b^2}}{a}$

Further we can state that $\sin\alpha = \dfrac{\sqrt{a^2 - b^2}}{a}$ since $ae = \sqrt{a^2 - b^2}$.

61

Therefore, we are able to gain a valuable insight into just what eccentricity (e) represents since $\sin\alpha = e$. *The angle α is 'angular eccentricity', the sine of which is eccentricity (e).*

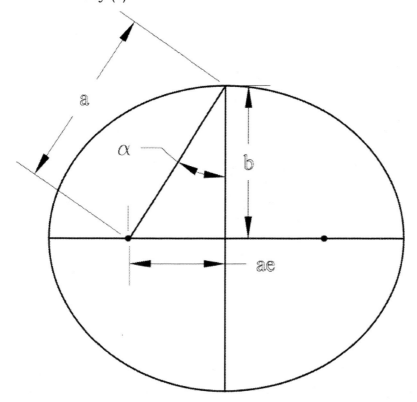

Figure 5.2 Angular Eccentricity.

Latitude

The flattened nature of the ellipse gives rise to three definitions for latitude. Unlike a circle, a normal to the ellipse (known as the *prime vertical*) does not pass through the origin of the ellipse. Figure 5.3 illustrates the three latitudes. *Geodetic latitude* (ϕ), is the included angle formed by the intersection of the ellipsoidal normal with the major axis (equatorial plane). *Geocentric latitude* (ψ), is the included angle formed by the intersection of a line extending from a point on the ellipse to the origin with the major axis. *Parametric* (or *reduced*) *latitude* (β), is the included angle formed by the intersection of a line extending from the projection of a point on the ellipse onto a concentric circle having a radius equal to the semi-major axis to the origin with the major axis.

The graticule for the ellipsoid consists of parallels identified by their geodetic latitude and meridians identified by their longitude. Geodetic

62

longitude (λ) is defined in the same manner as for the sphere. The ellipsoid becomes a more complex shape to analyze than the sphere due to its varying radii of curvature.

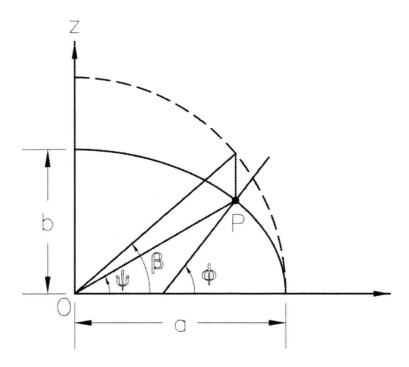

Figure 5.3 The three latitudes.

Ellipsoid of Revolution

A surface of revolution is defined as a surface that can be generated by revolving a plane curve about an axis in the plane (James and James, 1976). A simple surface of revolution is the sphere created by revolving a circle about a rotation axis. The ellipsoid of revolution is similar to the sphere in that all parallel circle planes (z = constant) form the trace of a circle. The ellipsoid of revolution differs from the sphere because the plane figure of an ellipse is rotated about the rotation axis instead of a circle resulting in meridian sections that trace ellipses.

One needs to carefully note that the term 'ellipsoid' is used interchangeably with the phase 'ellipsoid of revolution' in this text as shorthand notation. A strict definition of an ellipsoid may include three-dimensional figures more complex than that of the ellipsoid of revolution. For example, the triaxial ellipsoid explored in the past by geodesists as a more accurate model of the earth possesses an equator plane that traces the figure of an ellipse. This triaxial ellipsoid presents a computational surface that is much more difficult than the ellipsoid of revolution. We shall use the term 'ellipsoid' to refer to the simpler 'ellipsoid of revolution' in this text.

Two ellipsoid models have seen significant use in the United States. The student needs to be reminded that an ellipsoid is just a mathematical tool. Once the ellipsoid is developed it must be fitted to the earth using specific criteria. Ellipsoid fitting is discussed later in the material on geodetic datums. The defining parameters are presented for these two ellipsoids for the purpose of calculation:

Ellipsoid Name	Defining Parameters
Clarke Spheroid of 1866	a = 6,378,206.4 meters
	b = 6,356,583.8 meters
Geodetic Reference System of 1980	a = 6,378,137.000 meters
	1/f = 298.257222101.

Problem 5.1: Compute the following ellipsoid parameters for the Clarke Spheroid of 1866 (Clarke 1866): e, e², e', e'², and 1/f.

$$e = \frac{\sqrt{a^2 - b^2}}{a} = \frac{\sqrt{(6,378,206.4)^2 - (6,356,583.8)^2}}{6,378,206.4} = 0.0822718542261$$

$$e^2 = (0.0822718542261)^2 = 0.0067686579978$$

$$e' = \frac{\sqrt{a^2 - b^2}}{b} = \frac{(6,378,206.4)^2 - (6,356,583.8)^2}{6,356,583.8} = 0.082551710742$$

$$e'^2 = (0.082551710742)^2 = 0.00681478494643$$

$$f = 1 - \frac{b}{a} = 1 - \frac{6,356,583.8}{6,378,206.4} = 0.003390075304$$

$$1/f = \frac{1}{0.003390075304} = 294.978698208$$

Note: These ellipsoid parameters represent intermediate quantities used to compute other quantities such as geodetic azimuths, distances, and positions. Round-off error is a potential problem with these intermediate quantities. A calculator with at least twelve (12) places is recommended for these calculations.

Ewing, 1970 derives various formulas used to compute the parameters for the ellipsoid of revolution. It is instructive to state the derivation in terms of the radius of the parallel circle (p) and the perpendicular distance from the equatorial plane (z). Let the plane curve be the equation for an ellipse in the meridian plane:

$$\frac{p^2}{a^2} + \frac{z^2}{b^2} - 1 = 0 \ .$$

Figure 5.4 shows the geometry of the radius of the parallel (p) and the perpendicular distance (z) in the meridian section. It is convenient to derive the

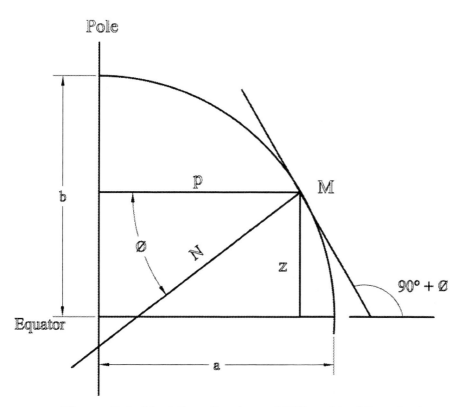

Figure 5.4 Meridian Section of Ellipsoid of Revolution.

formulas for p and z in terms of the following three variables: the semi-major axis (a), eccentricity (e), and geodetic latitude (φ). Note that the semi-major axis (a) and the eccentricity (e) are parameters associated with a particular ellipsoid.

65

Once the particular ellipsoid is chosen, these parameters can be considered constants. Therefore, the values of p and z for a particular place on the surface of the ellipsoid are a direct function of geodetic latitude (ϕ). The derivation for (p) and (z) of the meridian ellipse can be found in Appendix C. The formulas are stated as:

$$p = \frac{a\cos\phi}{\left(1 - e^2 \sin^2 \phi\right)^{1/2}}$$

$$z = \frac{a\left(1 - e^2\right)\sin\phi}{\left(1 - e^2 \sin^2 \phi\right)^{1/2}}$$

The formulas for (p) and (z) can be used for the mathematical construction of an ellipse. We will use these parameters (p) and (z) to derive formulas for the radii of curvature of the ellipsoid.

Radii of Curvature

Unlike the sphere, the ellipsoid does not possess a constant radius of curvature. We must consider the radii of curvature in two, mutually perpendicular planes. It will be seen that these principal radii of curvature are dependent on latitude only.

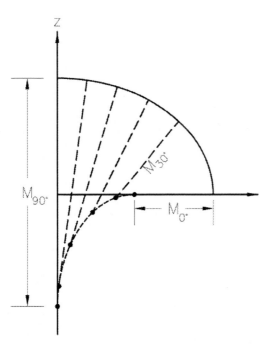

Figure 5.5 Radius of Curvature in the Meridian Section.

Meridian Plane

The radius of curvature in the meridian plane, *M*, is the radius of a circle which will only coincide with the shape of the meridian ellipse within a differential element because the radius is constantly changing along the meridian. As its name suggests, this radius is measured in the meridian plane. By inspection of figure 5.5 you can see that this radius differs for each point on the ellipse with latitude (from $\phi = 0°$ to $90°$).

Given a plane curve (the meridian ellipse) whose graph or shape is adequately captured by a twice-differentiable function. The formula for calculating the radius of curvature is given by Larson, Hostetler, and Edwards, 2002. The equation of the radius of curvature in terms of rectangular coordinates (*p*) and (*z*) is:

$$Radius\ of\ Curvature = \frac{\left(1 + \left(\frac{dz}{dp}\right)^2\right)^{\frac{3}{2}}}{\frac{d^2z}{dp^2}}$$

It remains to insert the first and second derivatives of the formula for the ellipse, in terms of the parameters (p) and (z), into the general equation of curvature to derive the radius of curvature in the meridian section (M). The resulting derivation of the equation for radius of curvature in the meridian section can be found in Appendix C. The radius of curvature in the meridian is stated as:

$$M = \frac{a\left(1 - e^2\right)}{\left(1 - e^2 \sin^2 \phi\right)^{3/2}}.$$

Prime Vertical

A normal section is created by passing a plane containing the normal to a particular point on the ellipsoid surface, such as point A in figure 5.6, through the ellipsoid surface. When this normal section has an azimuth of $90°$ (or $270°$) then its radius of curvature is defined as N, the radius of curvature in the prime vertical. Figure 5.7 shows the relationship between the lengths of the two principal radii. Note that N always extends from the minor axis to the ellipsoid surface and $N \geq M$.

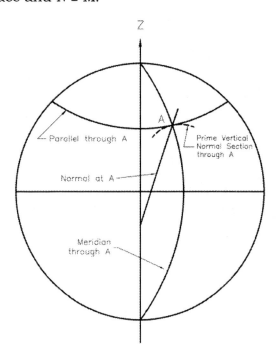

Figure 5.6 Radius of Curvature in the Prime Vertical.

The radius of curvature in the prime vertical can be expressed by utilizing the first derivative of our equation for the ellipse and the radius of the

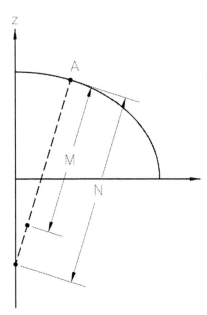

Figure 5.7 Radius of curvature in the meridian vs. prime vertical.

parallel. The slope of the tangent line at point A in figure 5.7 is the first derivative of z with respect to p. The first derivative can be used to find the angle (ϕ):

$$\phi = \left[\frac{-1}{\left(\dfrac{dz}{dp} \right)} \right] ;$$

The radius of the parallel circle, *p*, may be readily derived from the radius of curvature in the prime vertical and the geodetic latitude as shown in figure 5.4.

$$p = N \cos\phi .$$

$$N = \frac{p}{\cos\phi} .$$

To express the radius of curvature in the prime vertical as a function of geodetic latitude (ϕ) substitute the equation for the radius of the parallel circle for the variable (p).

$$N = \frac{a \cos\phi}{\cos\phi \left(1 - e^2 \sin^2\phi\right)^{\frac{1}{2}}} ,$$

$$\therefore \quad N = \frac{a}{\left(1 - e^2 \sin^2\phi\right)^{\frac{1}{2}}} .$$

Problem 5.2: Given the geodetic latitude of a place ϕ = 31° 48' 02", compute N, M, and p on Clarke 1866.

$$N = \frac{6,378206.4}{\left(1 - 0.0067686579978(\sin(31°48'02"))^2\right)^{\frac{1}{2}}} = 6,384,209.08909 \, meters.$$

$$M = \frac{6,378,206.4\left(1 - 0.0067686579978\right)}{\left(1 - 0.0067686579978(\sin(31°48'02"))^2\right)^{\frac{3}{2}}} = 6,352,937.51748 \, meters.$$

69

$$p = N \cos\phi = 6,384,209.08909 \cos(31°48'02'') = 5,425,860.0349 \text{ meters.}$$

We must remember that the radius of curvature of the prime vertical is contained in a special normal section—one that is oriented at 90° (or 270°) to the meridian. There are an infinite number of normal sections at a point on the surface of the ellipsoid. The radius of curvature of these sections is a function of azimuth. This radius of curvature of a normal section is given by:

$$R_\alpha = \frac{MN}{M \sin^2 \alpha + N \cos^2 \alpha}$$

The variable M is the radius of curvature of the meridian and the variable N is the radius of curvature in the prime vertical at a given geodetic latitude. The variable α is the azimuth of the normal section. Burkholder, 1987 lists the properties of the radius of curvature for normal sections as:

- $R\alpha$ = M for a normal section oriented at an azimuth of 0°.
- $R\alpha$ = N for a normal section oriented at an azimuth of 90°.
- $R_{30°} = R_{150°} = R_{210°} = R_{330°}$ due to symmetry.
- Values of $R\alpha$ will always be greater than M and smaller than N.

THREE TYPES OF LATITUDE (Revisited)

Derivation of the formulas to compute the three different types of latitude; *geodetic latitude* (ϕ), *geocentric latitude* (ψ), and *parametric* (or *reduced*) *latitude* (β), is given by Rapp, 1991 using the definition of parametric latitude (β) as shown in figure 5.8. The Pythagorean theorem can be used in right triangle $P_1 O P_2$ to state the relationship between the three sides.

$$(OP_2)^2 + (P_2 P_1)^2 = a^2 \text{ and given that } \frac{p^2}{a^2} + \frac{z^2}{b^2} - 1 = 0;$$

Since OP_2 = p and PP_2 = z, substitute these values in our equation for the ellipse;

$$\frac{(OP_2)^2}{a^2} + \frac{(PP_2)^2}{b^2} = 1,$$

Multiply through by a^2,

70

$$\left(OP_2\right)^2 + \frac{a^2}{b^2}\left(PP_2\right)^2 = a^2 = \left(OP_2\right)^2 + \left(P_2P_1\right)^2,$$

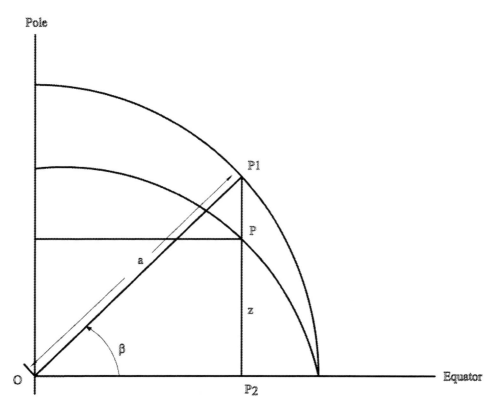

Figure 5.8   Parametric Latitude.

$$\frac{a^2}{b^2}\left(PP_2\right)^2 = \left(P_2P_1\right)^2,$$

$$PP_2 = z = \frac{b}{a}P_2P_1, \text{ so } z = \frac{b}{a}a\sin\beta \quad or \quad b\sin\beta,$$

$$\cos\beta = \frac{p}{a} \quad then \quad p = a\cos\beta.$$

Given the values for p and z computed with respect to the meridian ellipse:

$$\tan\psi = \frac{z}{p} = \frac{b\sin\beta}{a\cos\beta} = \frac{\dfrac{a(1-e^2)\sin\phi}{\left(1-e^2\sin^2\phi\right)^{1/2}}}{\dfrac{a\cos\phi}{\left(1-e^2\sin^2\phi\right)^{1/2}}} = \frac{b}{a}\tan\beta = \left(1-e^2\right)\tan\phi,$$

Thus $\tan \psi = \left(1 - e^2\right)^{\frac{1}{2}} \tan \beta = \left(1 - e^2\right) \tan \phi$.

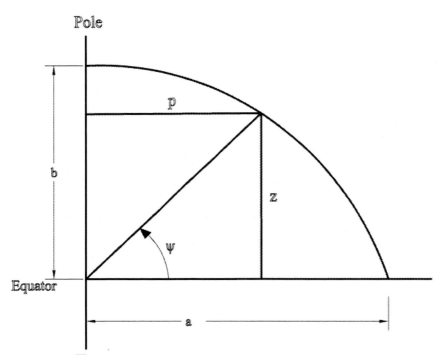

Figure 5.9 Geocentric latitude on the Ellipsoid.

Problem 5.3: Given the geodetic latitude of a place $\phi = 31° \, 48' \, 02''$ N, compute β, and ψ on Clark 1866.

$$\beta = \arctan\left(\left(1 - e^2\right)^{\frac{1}{2}} \tan \phi\right) = \arctan\left((1 - 0.0067686579978)^{\frac{1}{2}} \tan(31°48'02'')\right)$$

$$\beta = 31.713483° \quad or \quad 31° \, 42' \, 49'' \, N$$

$$\psi = \arctan\left(\left(1 - e^2\right) \tan \phi\right) = \arctan\left(\left(1 - 0.0067686579978\right) \tan\left(31°48'02''\right)\right)$$

$$\psi = 31.626542 \quad or \quad 31° \, 37'36'' \, N$$

Lengths of Arcs

Figure 5.10 shows a differential element of area, dA, on the surface of the ellipsoid. This element is bounded by arcs having differential lengths dG (along the meridian) and dL (along the parallel circle). The area of the differential element is given by $dA = dG \times dL$. The length of a differential arc along the parallel circle is $dL = pd\lambda_{radians} = N \cos \phi \, d\lambda_{radians}$.

72

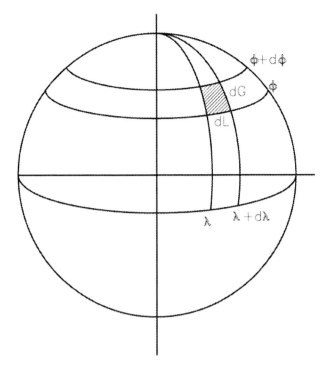

Figure 5.10 Differential element of Area on surface of Ellipsoid.

Since the latitude is constant along a parallel circle, we can compute the length of any arc along a parallel circle, L, from $L = N \cos\phi\, \Delta\lambda_{\; radians}$. Notice that convergence of the meridians creates the "cosine effect" whereby the arc length

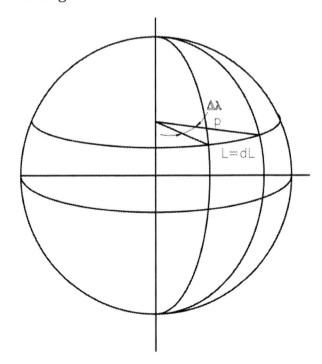

Figure 5.11 Length of a differential arc along a parallel circle.

along a parallel circle decreases from the equator to the pole for a given longitudinal difference. Figure 5.11 shows the geometry of the differential arc along a parallel. Note that one of the distinguishing features of an ellipsoid of revolution is that all parallels are circles. Each circle is successively smaller as the latitude increases, but each parallel still can be described as a circle divisible by 0° to 360° longitude.

The length of a differential arc along the meridian is given by $dG = M d\phi_{radians}$ (see figure 5.12). The length of any arc along a meridian can be computed by integration as follows:

$$G = \int_{\phi_1}^{\phi_2} M d\phi = a(1 - e^2) \int_{\phi_1}^{\phi_2} \frac{d\phi}{(1 - e^2 \sin^2 \phi)^{3/2}}$$

This form is an *elliptic integral* that has no closed form solution. A series expansion may be used to estimate the solution. The following series may be used to compute a meridian arc length, S, from the equator to a point at latitude ϕ.

$$S_\phi = \frac{a}{1+n} [a_0 \phi - a_2 \sin 2\phi + a_4 \sin 4\phi - a_6 \sin 6\phi + a_8 \sin 8\phi]$$

where:

$$n = \frac{f}{2-f}$$

$$a_0 = 1 + \frac{n^2}{4} + \frac{n^4}{64}$$

$$a_2 = \frac{3}{2}\left(n - \frac{n^3}{8}\right)$$

$$a_4 = \frac{15}{16}\left(n^2 - \frac{n^4}{4}\right)$$

$$a_6 = \frac{35}{48} n^3$$

$$a_8 = \frac{315}{512} n^4$$

Accordingly, $G = S_{\phi_2} - S_{\phi_1}$.

A special arc length is given by the Quadrant of the Meridian. The Quadrant of the Meridian is the meridian arc length from the equator to the pole. The series

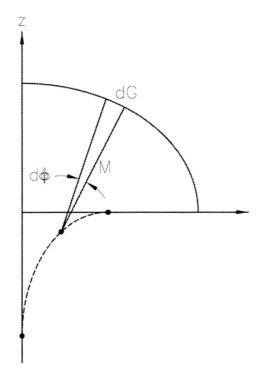

Figure 5.12 Length of Differential Arc along the meridian.

expansion is simplified when the geodetic latitude ϕ = 90° since the sine of 180° is zero. The simplified formula for the Quadrant of the Meridian is:

$$S_\phi = \frac{a}{1+n}\left(a_0\phi_{\;radians}\right)$$

The Quadrant of the Meridian holds an interesting historical perspective. Early geodesists performed a great number of arc measuring projects on the surface of the earth to determine the earth's shape. Delambre and Méchain measured a meridian arc between Dunkirk and Montjouy, France upon which the foundation of the metric system was based. The meter was set as one ten-millionth of the Quadrant of the Meridian.

Problem 5.4: Compute the Quadrant of the Meridian for Clarke 1866 ellipsoid.

$$n = \frac{f}{2-f} = \frac{0.003390075304}{2-0.003390075304} = 0.00169791568301$$

$$a_0 = 1 + \frac{n^2}{4} + \frac{n^4}{64} = 1 + \frac{(0.00169791568301)^2}{4} + \frac{(0.00169791568301)^4}{64}$$

$$a_0 = 1.00000072073$$

$$S_{0° \to 90°} = \frac{6,378,206.4}{1 + 0.00169791568301}(1.00000072073(1.57079632679))$$

$$S_{0° \to 90°} = 10,001,888.043 \; meters.$$

Area

The area bounded by two meridians and two parallel circles may be derived:

$$dA = dG \times dL = MN \cos\phi \, d\phi \, d\lambda$$

$$A = \int dA = \int_{\phi_1}^{\phi_2} \int_{\lambda_1}^{\lambda_2} MN \cos\phi \, d\phi \, d\lambda$$

$$A = \frac{(\lambda_2 - \lambda_1)a^2(1 - e^2)}{2}\left[\frac{\sin\phi}{1 - e^2\sin^2\phi} + \frac{1}{2e}\ln\left(\frac{1 + e\sin\phi}{1 - e\sin\phi}\right)\right]_{\phi_1}^{\phi_2}$$

Problem 5.5: Compute the area on the surface of the Clarke 1866 ellipsoid bounded by ϕ = 30°00'00"N, ϕ = 30°03'00"N, λ = 75°00'00"W, and λ = 75°02'00"W.

This problem requires evaluation of the right quantity of the equation at both latitudinal values and determining the difference between these two quantities before multiplying by the left quantity. The right quantity evaluated at ϕ = 30°00'00"N:

$$\left[\frac{\sin 30°00'00"}{1 - (0.0067686579978)\sin^2 30°00'00"} + \frac{1}{2(0.0822718542261)}\ln\left(\frac{1 + (0.0822718542261)\sin 30°00'00"}{1 - (0.0822718542261)\sin 30°00'00"}\right)\right]$$

= 1.0011298305

The right quantity evaluated at ϕ = 30°03'00"N:

$$\left[\frac{\sin 30°03'00"}{1 - (0.0067686579978)\sin^2 30°03'00"} + \frac{1}{2(0.0822718542261)}\ln\left(\frac{1 + (0.0822718542261)\sin 30°03'00"}{1 - (0.0822718542261)\sin 30°03'00"}\right)\right]$$

=1.0026460839

76

The left quantity:

$$\frac{\left[-75°02'00''-(-75°00'00'')\right]\left(\pi\big/_{180°}\right)(6,378,206.4\,\text{m})^2\,(1-0.0067686579978)}{2}$$

= -11,753,674,805 m²

Area = -11,753,674,805 m² (1.0026460839 – 1.0011298305) = -17,821,550 m²

  = 17,821,550 m² (negative area impossible)

The Geodesic

The *geodesic* (or *geodetic line*) is analogous to the great circle on the sphere in that it represents the shortest distance between two points on the surface of the ellipsoid. The geodesic possesses double curvature as shown in figure 5.13. The intersection of the ellipsoid and the normal section at point A that contains point B forms the bottom curve in the figure. The intersection of the ellipsoid and the normal section at point B that contains point A forms the top curve in the figure. The geodesic lies between these two curves as shown. Note that the normal sections would be common if $\phi_A = \phi_B$.

Geodesic actually is a generic term representing the shortest surface distance between any two points lying on the same surface. Therefore, the geodesic on a plane is a straight line while the geodesic on a sphere is a great

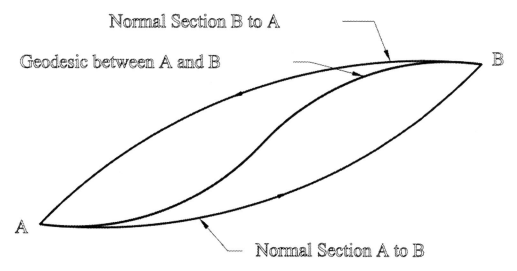

Figure 5.13 Shortest distance between two points on the ellipsoid (geodesic].

circle arc. The geodesic on the ellipsoid must satisfy *Clairaut's Theorem*: N $\cos \phi$ $\sin \alpha = constant$, where (α) is the geodetic azimuth.

Computations on the Ellipsoid Surface

It may be desirable to compute the geodetic coordinates for a point given the geodetic coordinates of one point and the geodetic length (length of the geodesic) and azimuth (of the geodesic) at the known point to the unknown point. This is known as the *direct solution* or *direct geodetic problem*. The *inverse solution* or *inverse geodetic problem* involves determination of the geodetic length and azimuths given geodetic coordinates of two points. Given the complex nature of the ellipsoid surface, many solutions to these problems provide acceptable accuracies only for lines of limited length.

The formulas presented below for the solution of the geodetic direct and inverse were developed by Bowring (1981) and are intended for use on lines of 150 km or less. The equations are derived from an orthomorphic projection of the ellipsoid onto a sphere. Presented without derivation, readers are encouraged to read the source paper for more information. Readers are referred to the literature for additional solutions to these problems.

Common Equations

The following equations are labeled the *common equations* by Bowring. They are used in both the direct and inverse solution:

$$A = \left(1 + e'^2 \cos^4 \phi_1\right)^{1/2}$$

$$B = \left(1 + e'^2 \cos^2 \phi_1\right)^{1/2}$$

$$C = \left(1 + e'^2\right)^{1/2}$$

$$w = \frac{A\left(\lambda_2 - \lambda_1\right)}{2}$$

Direct Solution

Given: the geodetic coordinates of a point (λ_1, ϕ_1), the geodetic distance from this point to a second point (s), and the geodetic azimuth from the first point to the second point $(a_{1\rightarrow 2})$.

Find: the geodetic coordinates of the second point (λ_2, ϕ_2) and the geodetic azimuth from the second point to the first point $(a_{2\rightarrow 1})$.

$$\sigma = \frac{sB^2}{aC} \; ; \qquad \lambda_2 = \lambda_1 + \frac{1}{A}\arctan\left(\frac{A\tan\sigma\sin\alpha_{1\rightarrow 2}}{B\cos\phi_1 - \tan\sigma\sin\phi_1\cos\alpha_{1\rightarrow 2}}\right)$$

$$D = \frac{1}{2}\arcsin\left(\sin\sigma\left(\cos\alpha_{1\rightarrow 2} - \frac{1}{A}\sin\phi_1\sin\alpha_{1\rightarrow 2}\tan w\right)\right)$$

$$\phi_2 = \phi_1 + 2D\left(B - \frac{3}{2}e'^2 D\sin\left(2\phi_1 + \frac{4}{3}BD\right)\right)$$

$$\alpha_{2\rightarrow 1} = \arctan\left(\frac{-B\sin\alpha_{1\rightarrow 2}}{\cos\sigma\left(\tan\sigma\tan\phi_1 - B\cos\alpha_{1\rightarrow 2}\right)}\right)$$

Inverse Solution

Given: the geodetic coordinates of two points (λ_1, ϕ_1) and (λ_2, ϕ_2).

Find: the geodetic distance between the two points (s), the geodetic azimuth from the first point to the second point $(a_{1\rightarrow 2})$ and the geodetic azimuth from the second point to the first point $(a_{2\rightarrow 1})$.

$$\Delta\phi = \phi_2 - \phi_1 \; ; \qquad D = \frac{\Delta\phi}{2B}\left(1 + \frac{3e'^2}{4B^2}\Delta\phi\sin\left(2\phi_1 + \frac{2}{3}\Delta\phi\right)\right) \; ; \qquad E = \sin D\cos w$$

$$F = \frac{1}{A}\sin w\left(B\cos\phi_1\cos D - \sin\phi_1\sin D\right); \quad \tan G = \frac{F}{E}$$

$$\sin\frac{\sigma}{2} = \left(E^2 + F^2\right)^{1/2}; \quad \tan H = \left(\frac{1}{A}\left(\sin\phi_1 + B\cos\phi_1\tan D\right)\tan w\right)$$

$$\alpha_{1\rightarrow 2} = G - H \; ; \qquad \alpha_{2\rightarrow 1} = G + H \pm 180° \; ; \qquad s = \frac{aC\sigma}{B^2}$$

Problem 5.6: Given a point having geodetic coordinates 40°00'00.0000"N latitude and 120°00'00.0000"W longitude based upon the Clarke 1866 spheroid. The geodetic distance to a second point is 10,000.000 m at an azimuth of 45°00'00.00". Determine the geodetic coordinates of the second point.

First, compute the common equation values:

$$A = \left(1 + 0.00681478494643 \cos^4 40°00'00.0000"\right)^{\frac{1}{2}} = 1.0011726906271$$

$$B = \left(1 + 0.00681478494643 \cos^2 40°00'00.0000"\right)^{\frac{1}{2}} = 1.00199754489045$$

$$C = \left(1 + 0.00681478494643\right)^{\frac{1}{2}} = 1.00340160700784$$

Then compute the direct solution. In the interest of saving space, variables may be shown below rather than their values:

$$\sigma = \frac{10,000.000\,\text{m}\,(B)^2}{6,378,206.4\,\text{m}\,(C)} = 0.00156877244907 \text{ radians}$$

$$\lambda_2 = -120°\left(\frac{\pi}{180°}\right) + \frac{1}{A}\arctan\left(\frac{A\tan\sigma\sin 45°}{B\cos 40° - \tan\sigma\sin 40°\cos 45°}\right)$$

$$\lambda_2 = -2.09294857051253 \text{ radians}$$

λ_2 = -119.91711983° = -119°55' 01.6314" = 119°55'01.6314" W

$$w = \frac{A\left[-119°55'01.6314" - \left(-120°00'00.0000"\right)\right]}{2} = 0.00072411410752 \text{ radians}$$

$$D = \frac{1}{2}\arcsin\left[\sin\sigma\left(\cos 45° - \frac{1}{A}\sin 40°\sin 45°\tan w\right)\right] = 0.00055438684667$$

$$\phi_2 = 40°\left(\frac{\pi}{180°}\right) + 2D\left(B - \frac{3}{2}e'^2 D\sin\left(2\phi_1 + \frac{4}{3}BD\right)\right) = 0.699242683127 \text{ radians}$$

0.699242683127 *radians* = 40.0636545986° = 40°03'49.1566" N

$$\alpha_{2\to1} = \arctan\left[\frac{-B\sin 45°}{\cos\sigma\left(\tan\sigma\tan 40° - B\cos 45°\right)}\right] = \arctan\left(\frac{-0.70851925872}{-0.70720203103}\right)$$

$\alpha_{2\to1}$ = 45.05330962° + 180° (3rd quadrant) = 225.05330962° = 225°03' 11.91"

Problem 5.7: Determine the geodetic azimuths and distance between two points on the Clarke spheroid of 1866 having the following coordinates: point 1 = (42°15' 00.0000"N, 121°45' 00.0000"W) and point 2 = (42°00' 00.0000"N, 121°30' 00.0000"W).

Compute the common equations:

$$A = \left(1 + 0.00681478494643 \cos^4 42.25°\right)^{1/2} = 1.00102244289596$$

$$B = \left(1 + 0.00681478494643 \cos^2 42.25°\right)^{1/2} = 1.00186524870796$$

$$C = \left(1 + 0.00681478494643\right)^{1/2} = 1.00340160700784$$

$$w = \frac{A\left[-121.50° - (-121.75°)\right]}{2} = 0.00218389218936 \text{ radians}$$

Now compute the inverse solution:

$$\Delta\phi = 42.00° - 42.25° = -0.25° = -0.00436332312999 \text{ radians}$$

$$D = \left(\frac{\Delta\phi}{2B}\right)\left[1 + \frac{3\left(0.00681478494643\right)}{4 B^2}\left(\Delta\phi\right)\sin\left(2\left(42.25°\right) + \frac{2}{3}\left(\Delta\phi\right)\right)\right]$$

$$D = -0.0021775516535 \text{ radians}$$

$$E = \sin D \cos w = -0.0021775447398$$

$$F = \left(\frac{1}{A}\right)\sin w \left(B \cos 42.25° \cos D - \sin 42.25° \sin D\right) = 0.0016211067118$$

$$G = \arctan\left(\frac{F}{E}\right) = \arctan\left(\frac{0.0016211067118}{-0.0021775447398}\right) = -36.66640794° + 180° \text{ (2nd quadrant)}$$
$$= 143.33359206° = 143°20'00.93"$$

$$\sigma = 2\left[\arcsin\left((E^2 + F^2)^{1/2}\right)\right] = 0.0054294405491 \text{ radians}$$

$$H = \arctan\left[\frac{\left(\sin 42.25° + B \cos 42.25° \tan D\right)\tan w}{A}\right] = \arctan\left(\frac{0.00146485224}{1.00102244289596}\right)$$
$$= 0.08384407° \text{ (1st quadrant)} = 0°05'01.84"$$

$$\alpha_{1\to2} = G - H = 143°20'00.93" - 0°05'01.84" = 143°14'59.09"$$

$$\alpha_{2\to1} = G + H \pm 180° = 143°20'00.93" + 0°05'01.84" + 180° = 323°25'02.77"$$

$$s = \frac{6{,}378{,}206.4 \, \text{m} \, (C) \, \sigma}{B^2} = 34{,}618.625 \, \text{m}$$

Study Questions

1. Prove that $(1-e^2)(1+e'^2) = 1$.
2. The Bessel ellipsoid was defined as: a = 6, 377, 397 meters (exact) and the reciprocal flattening of 299.2 (exact). Compute R_α for a north azimuth of 62°50'37" at a geodetic latitude of 31° 48' 02". What does this value represent?
3. Given the defining parameters of the GRS 80 ellipsoid: a = 6,378,137.000 meters; 1/f = 298.257222101.
   a) Compute in steps of 10°, starting at a geodetic latitude of ϕ = 0°, and ending at ϕ = 90°, the following quantities: the spherical latitude (ψ), the difference between the spherical and geodetic (ψ-ϕ), the radius of curvature in the meridian plan (M), and the radius of curvature in the prime vertical (N). Present your computed values in a table such as the one below:

ϕ deg	ψ dms	ψ-ϕ dms	M km	N km
00				
10				
20				
30				
40				
50				
60				
70				
80				
90				

   b) Prepare two graphs based on the data in your table in a) above. Place geodetic latitude on the vertical axis in both graphs. The first graph is a plot of geodetic latitude to the difference (ψ-ϕ) at 10-degree increments. (ψ-ϕ) should be placed on the horizontal axis. The second graph is a plot of N and M as a function of geodetic latitude. The range of computed values of M and N in kilometers should be placed on the horizontal axis.
   c) Consider the results of your work. What can you conclude about the difference (ψ-ϕ) based on the table and graph? What can you conclude about the behavior of M and N at different latitudes?
4. Consider Problems 5.1, 5.2, 5.3, and 5.4 in this chapter. Recompute the answers using the GRS 80 ellipsoid.
5. Given a rectangular coordinate system scaled in inches. The foci of an ellipse are located at (-2, 0) and (2, 0). Use the string method to construct an ellipse with a semi-major axis of 4 inches.

   Construction steps:
   - Plot the foci.

- Total string length should be twice the semi-major axis. Anchor both ends of the string at the foci.
- Trace ½ of the ellipse in one direction.
- Trace the other ½ of the ellipse in the other direction.

   a) Compute the parameters of your ellipse: $1/f$, b, e^2, and e'^2.

   b) Locate a point P at latitude of 35°. Compute N, M, p, and z. The longitude for point P is not given. Explain why longitude is not necessary to answer this question.

6. An ellipsoid has a semi-major axis of 6,378,399.922 meters and a semi-minor axis of 6,356,912.946 meters. Calculate the square of the eccentricity (e^2) and the reciprocal of the flattening ($1/f$) for this ellipsoid.

7. At what latitude are the radius of curvature in the meridian and the radius of curvature in the prime vertical equal? Mathematically support your answer.

8. Compute the instantaneous radius of curvature (R_a) on the GRS 80 ellipsoid at a latitude of 42° 15'16" for azimuths of 0°, 10°, 20°, 30°, 40°, 50°, 60°, 70°, 80°, and 90°. Place the values in a simple table. Compute the radius of curvature in the prime vertical (N). Does the value of (N) or (a) agree with any of the values in your table? If so, explain why.

9. Given that flattening on the ellipsoid can be calculated as $f = \dfrac{a-b}{a}$.

   Flattening on the ellipsoid can vary from a value just greater than zero, but less than one (1). Use the formula to determine the type of geometric figure represented by:

   a) f = 1

   b) f = 0.

10. You have created your own ellipsoid with a semi major axis of 6,378,222.000 meters and a reciprocal flattening of 334.0000000.

   a) Compute the ellipsoid parameters: f, b, e^2, and e'^2.

   b) Compute the N, M, and p at the latitude of 43° 27'39".

CHAPTER SIX

GEODETIC PERSPECTIVE ON THE USPLSS

From actual practice in surveying it is evident that the meridians do converge in going to the north, and, therefore, at such distances on the meridians as will show this convergency in practical surveying, there should be another set of exact miles measured, from which to carry on the surveys farther to the north.
Letter from the Surveyor General of Mississippi to the Commissioner of General Land Office in 1821.

The practical use of convergence in geodesy lies in the fact that the forward and back azimuths of a line do not differ by 180°, as is the case in plane surveying, but by 180° plus the convergence.
Ewing and Mitchell, 1970.

The study of geodesy in the United States is not complete without mention of one of the largest survey projects on record--the U.S. Public Land Survey System commonly known as the Rectangular System. The so-called Rectangular System is really not rectangular at all due to the convergence of the meridians.

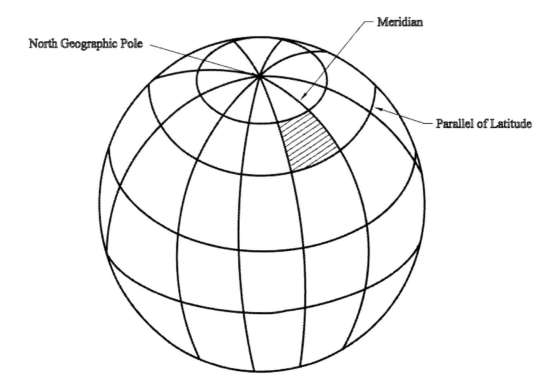

Figure 6.1 The basic shape of a "square" as defined under the USPLSS.

The Land Ordinance of 1785 authorized the U.S Public Land Survey System (USPLSS). This law enacted by Congress stated:

> "The surveyors...shall proceed to divide the said territory into townships of six miles square, by lines running due north and south, and others crossing them at right angles...".

Consideration of the law brings the conclusion that since the meridians converge toward the North Pole, townships exactly 6 miles square are an impossibility!

The implementation of "lines running due north and south" result in lines oriented to true north determined by astronomic observations. Later instrumentation of the era consisted of the Bert Solar Compass and the transit with solar attachment. Astronomic observations for azimuth were conducted using the *altitude method*. The altitude method solves the *PZS triangle* (a spherical triangle on the celestial sphere containing the observer's zenith, the sun, and the celestial pole as vertices) using the observer's known latitude, the sun's declination, and sun's altitude at the time of observation. Anderson and Mikhail, 1998 give the probable standard deviation of an azimuth obtained by the altitude method at ± 10-15 seconds. The altitude method was favored by deputy surveyors during implementation of the USPLSS because precise time is not required for this method.

A second consideration is that the instrumentation used by the deputy surveyors was oriented to the plumb line (gravity vector). Unlike today's surveying instruments (e.g., total station) that make use of optical plummets for plumbing the instrument over the point, older instruments (transits, etc.) made use of an actual plumb bob and string—the plumb line. The plumb line was attached to the bottom of the instrument and used to orient the instrument over the desired point on the topographic surface. The plumb line was aligned with the gravity vector due to the mass of the plumb bob and gravity acceleration. Then the surveyor used the action of gravity on sensitive vial level bubbles to establish a horizontal plane perpendicular to the plumb line. Therefore, the vertical axis of the instrument is oriented to the direction of the plumb line. Even though few surveyors set an instrument over a point with a plumb bob today, the modern total station is still leveled up to the plumb line.

Kissam, 1956 states:

An astronomic observation that is used for surveying purposes basically determines two things as follows:
1) The coordinates of the point on the celestial sphere where the local direction of gravity strikes it. It must be remembered that when an instrument is leveled, the vertical axis is made to coincide with the direction of gravity. The zenith, therefore, is the local direction of gravity projected to the celestial sphere. The measurements made and the trigonometry applied give the celestial coordinates of the zenith. These are translated to the earth coordinates of latitude and longitude.
2) The azimuth of a mark measured in a plane perpendicular to gravity. Thus, latitude, longitude, and azimuth, determined astronomically depend on the local direction of gravity and hence are referred to the geoid.

Latitude, longitude, and azimuth determined astronomically are denoted as Φ, Λ, and A, respectively. An important point is that the layout of the Rectangular System involved astronomic observations. These astronomic observations were not reduced to a reference ellipsoid to obtain geodetic latitude (ϕ), geodetic longitude (λ), or geodetic azimuth (α). It is safe to say that the precision lost in the layout of the Rectangular System by the use of astronomic observations was not significant in light of the general precision of the instrumentation used and survey practice implemented by the deputy surveyors.

Further, one can conclude that the only way to cross two meridians at right angles is on an arc or parallel of latitude. The requirement of the law that east west lines cross at right angles introduces the concept of the rhumb line. Dutton, 1943 defines the rhumb line as a line on the earth's surface that intersects all meridians at the same angle, i.e., a line of constant azimuth. Parallels of latitude are special rhumb lines which meet each meridian at right angles and remain equidistant from the poles. All other rhumb lines are loxodromic curves. These rhumb lines act as spirals that curve in approach to one of the poles, but never reach the pole. The special rhumb lines (parallels of latitude) are necessarily curved lines on the ground because of the convergence of the meridians.

Since the township was defined by law to be two converging straight lines forming the east and west boundaries, it is not square. The north and south boundaries were to coincide with parallels of latitude where located along standard parallels. The north boundary is shorter than the south boundary due to convergence. The basic shape of this geometric figure is shown in exaggerated fashion in figure 6.1.

The authorization of the USPLSS by law and its subsequent implementation in 30 of the 50 states provides a massive survey project with geodetic implications for study by practicing professional surveyors today. It is clear that the intent of the USPLSS was never to reduce measurements to the ellipsoid of revolution as a reference surface. However the implementation of the USPLSS required application of geodetic theory and resulted in the institution of the quadrangle (correction lines). To dismiss geodesy as irrelevant to the practice of land surveying in the United States is to ignore the lessons of history. Now that the USPLSS is essentially complete and aging, one needs to apply the fundamentals of geodesy to understand what was accomplished.

Since convergence is at the heart of the problem of placing a rectangular system over a large area, this chapter will derive several formulas to quantify convergence as a function of latitude, consider the impact of convergence on azimuth and bearing, and then finish with further analysis of the effect of convergence on the USPLSS.

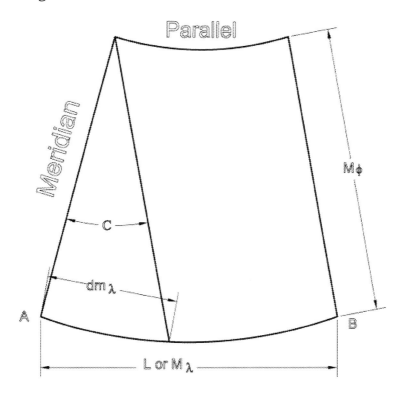

Figure 6.2 Quantifying convergence.

Definition of Convergence

Convergence does not exist where meridians are parallel. Parallelism is an assumption in the use of the tangent plane as a reference surface that allows the use of plane trigonometry for calculations. Of course, this assumption is never true. *Angular convergence* (C) is the measure of the degree of non-parallelism experienced by two meridians spaced a linear distance (m_λ) apart at a mean latitude (ϕ_m). *Linear convergence* (dm_λ) is the measure of the degree of non-parallelism experienced by two meridians spaced a linear distance (m_λ) apart, having a meridional length (m_ϕ), at a mean latitude (ϕ_m). These relationships are shown in figure 6.2.

Convergence on the Sphere

A formula to quantify angular convergence, in terms of the angular distance the meridians are apart $(\Delta\lambda)$ and their mean latitude (ψ_m), can be derived on the sphere.

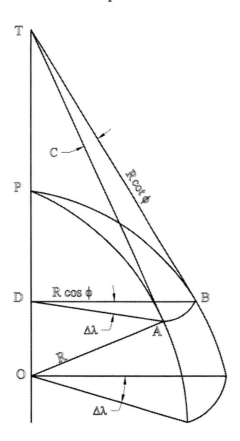

Figure 6.3 Use of the extended pole to quantify convergence on the sphere.

The extended pole (T) is the intersection of three lines; two of which are tangent to the meridians at points A and B and the third is coincident with the rotation axis of the sphere. Angle C of triangle ABT is the angular measure of convergence as shown in figure 6.3.

Express the arc length AB within triangle ABD as $AB = (R\cos\psi)\Delta\lambda$. Consider triangle AOT. Since the angle AOT is (90°- ψ), then angle OTA equals ψ by the definition of latitude and the fact that the sum of the angles in a plane triangle is 180°. (Remember, line AT is tangent to the sphere at A).

$$\cot\psi = \frac{AT}{R},$$

$$AT = R\cot\psi.$$

Now express the arc length AB within triangle ATB as $AB = (R\cot\psi)C$. Eliminate AB by setting the equations equal:

$$(R\cos\psi)\Delta\lambda = (R\cot\psi)C,$$

$$C = \frac{R\cos\psi}{R\cot\psi}\Delta\lambda,$$

$$C = \Delta\lambda\,\sin\psi. \tag{6.1}$$

Equation 6.1 is in terms of spherical latitude and longitude. It is convenient to use the formula for arc length (AB) within triangle ABD to eliminate the change in longitude ($\Delta\lambda$) from equation 6.1:

$$\Delta\lambda = \frac{AB}{R\cos\psi} = \frac{m_\lambda}{R\cos\psi},$$

$$C = \frac{m_\lambda}{R\cos\psi}\sin\psi = \frac{m_\lambda\tan\psi}{R}.$$

m_λ is the distance measured on the parallel shown in figure 6.2.

Problem 6.1: A linear traverse is primarily of east-west extent (see figure 6.4). The difference in departure of the known starting point and known ending point is 10,362.78 feet. The mean latitude of the traverse is 31° 48' 02."

    a) What is the expected angular misclosure if the crew chief can measure each horizontal angle at a precision of ±15" and the traverse can be completed with 8 horizontal angles?

$$\sigma_{Angular\ Misclosure} = \sqrt{(15")^2 \times 8} = 42"$$

b) If the actual angular misclosure is 00° 01' 32" should the crew chief look for another job? No, the crew chief should suspect that another factor other than a blunder could cause such an apparent discrepancy. That factor is the convergence of the meridians.

c) How much of the 00° 01' 32" computed angular misclosure is due to the convergence of the meridians.

$$C = \frac{10{,}362.78\ feet \times \tan 31°48'02"}{6{,}371{,}000\ meters \times 1\ foot \Big/ 0.3048\ meters} = 0.000307399479894\ radians$$

$$C = 0.000307399479894\ radians \times 180° \Big/ \pi = 0.01761269°\ or\ 00°01'\ 03.4"$$

Convergence is a *systematic error* because it follows a defined pattern that can be mathematically quantified. This systematic error can be calculated and removed from the expected pool of random error to be allocated in the traverse adjustment (See figure 6.4). The disposition of the systematic error caused by convergence is a larger issue. It is clear that convergence is not random and should not be allocated by the compass rule or by the use of a least squares adjustment.

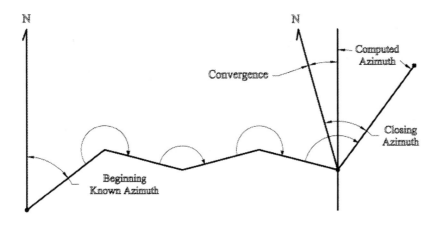

Figure 6.4 Convergence as systematic error in a traverse.

Section 2-74 of the Manual of Surveying Instructions 1973 suggests the allocation of convergence between stations based on departure independent of the allocation of random error. Clearly, the context of this discussion is on the tangent plane as the geometric reference surface rather than on the sphere or ellipsoid. On the surface of the ellipsoid, it is possible to precisely compute the geodetic position of each station using numerical integration. This computation, known as the geodetic direct, (see Chapter 5) makes allowance for the effects of convergence.

The use of the tangent plane as a geometric reference surface requires that the systematic error be allocated on some basis to allow closure of the traverse. The 1973 Manual states:

> ...in the subdivision of large areas as in the rectangular survey of the public lands, more stress should be placed on accuracy of distance measurement if those values are to be as good as the values required in the direction of lines.

It must be remembered that before the widespread use of the Electronic Distance Measuring Instrument (EDMI) and the Global Positioning System (GPS), directions could be measured much more precisely than distances. The instructions of the several Surveyor Generals to the deputy surveyors repeatedly addressed the myriad of problems associated with obtaining consistent measurements from a chaining crew. Considering the quality of the measurements that could be made given the working conditions and instrumentation, the added precision obtainable by using an ellipsoid as a reference surface rather than a tangent plane was never required.

Convergence on the Ellipsoid

Using an ellipsoid of revolution as a reference surface rather than a sphere can result in a more precise estimate of convergence. The properties of the ellipsoid can be introduced into the formula by substituting N for R. N is the symbol for the radius of curvature in the prime vertical developed in Chapter 5. It is also necessary to substitute geodetic latitude (ϕ) for spherical latitude (ψ).

$$C = \frac{m_\lambda \tan \phi}{\dfrac{a}{\sqrt{1 - e^2 \sin^2 \phi}}} = \frac{m_\lambda \tan \phi \sqrt{1 - e^2 \sin^2 \phi}}{a}$$

Problem 6.2: Recompute the convergence of the meridians in Problem 1 using the GRS80 ellipsoid of revolution instead of the sphere as a geometric reference surface.

$$C = \frac{10,362.78 \; feet \; \tan 31°48'02'' (1 - 0.0066943800229(\sin 31°48'02'')^2)^{1/2}}{6,378,137 \; meters \times {}^{1 \, foot}\!\!\big/\!_{0.3048}}$$

$$C = 0.00030676997019 \; radians \times {}^{180°}\!\!\big/\!_{\pi} = 0.0175766 \; or \; 00°01'\,03.3''.$$

The computation of linear convergence is quite simple. Apply the basic formula for the length of an arc, s = rθ. In figure 6.2, s = dm_λ (linear convergence), r = m_ϕ (the measurement along the meridian), θ is the angular convergence (*C*), and m_λ is the measurement along the parallel. This gives the formula for linear convergence as stated in Section 2-79 of the 1973 Manual:

$$dm_\lambda = \frac{m_\lambda m_\phi}{a} \tan \phi \; \sqrt{1 - e^2 \sin^2 \phi} \; .$$

Problem 6.3: Compute the linear convergence in Problem 6.2. Assume that $m_\phi = m_\lambda$.

$$dm_\lambda = \frac{(10,362.78 \; feet)^2}{6,378,137m \times {}^{1 \, foot}\!\!\big/\!_{0.3048m}} \tan 31°48'02'' \sqrt{1 - 0.0066943800229(\sin 31°48'02'')^2}$$

$$dm_\lambda = 3.18 \; feet.$$

Note: In the setup of these problems it is important that all distance measurements be in the same units.

Mean Bearing and Azimuth

To accomplish traverse computations on the tangent plane, the student learns to add 180° to the forward azimuth of a station to obtain the back azimuth at the next station. This assumption is incorrect due to convergence of the meridians and will result in significant systematic error for a plane survey traverse of large extent. This was illustrated in Problem 6.1.

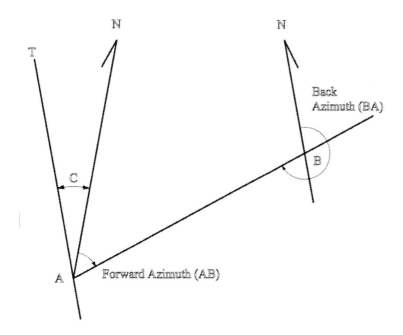

Figure 6.5 A line does not cross meridians at the same angle.

The difference between a forward and back azimuth is $180°$ plus the convergence angle between the two meridians from which the azimuths are calculated. In figure 6.5, reference line AT is parallel to the meridian through point B. Therefore, the angle between line AT and the meridian through point A is the convergence between the two meridians. The relationship between the back azimuth and the forward azimuth is:

$$Back\ Azimuth = Forward\ Azimuth + 180° + Convergence\ (C)$$

This explains why error can be introduced into a plane survey of large extent. The fact that convergence is a function of area is evident in the formula for linear convergence expressed in the term ($m_\lambda m_\phi / a$). To avoid the accumulation of error due to the convergence of the meridians, it is necessary that the areal extent of plane surveys be minimized.

Section 2-74 of the 1973 Manual states that:

> When computing latitudes and departures and transferring a geographic position by
> means of a long connecting line, the mean azimuth should be employed for the direction
> of the line, i.e. ---the mean between the forward azimuth and the back azimuth ± 180°.
> That azimuth or bearing angle will be the direction of the chord of the great circle that
> passes through the ends of the connecting line.

Wahl and Hintz, 1996 bring to our attention that this frame of reference is not
orthogonal because the reference meridians are not parallel. They state:

> This mean bearing is essentially identical to the bearing of the straight traverse line with
> reference to its midpoint. Thus the point of record for determining the bearing of a
> traverse line can be said to be the meridian of the midpoint of the line.

Again, we must be reminded that the context of this discussion is on
calculations made on the tangent plane as the reference surface rather than on
the ellipsoid. As stated previously, on the surface of the ellipsoid, it is possible
to precisely compute the geodetic position of each station using numerical
integration without the use of mean bearings.

Distances

The reduction of slope distances to the horizontal using the principles of
plane trigonometry is a practice of surveyors on the tangent plane. Anderson
and Mikhail,1998 define a short line as one less than two miles long or one with
a mean zenith angle between 85° and 95°. A series of such short lines reduced
to the horizontal using plane trigonometry will follow the curvature of the earth.
The aggregate of short horizontal lines can be further reduced to the ellipsoid
surface to form a geodetic line. The reduction of long lines to the ellipsoid
surface is treated in Chapter Seven *Reduction of Observations*. This step was
not taken in the implementation of the USPLSS.

Geodetic Implications

To bring this chapter to a close, we can list several geodetic implications
due to the law defining the USPLS and the survey practice used to implement
the system.

1. Measured distances were reduced to the horizontal on a set of small
   tangent planes that effectively follow the earth's topography. The next

possible step to reduce these horizontal distances to the surface of a reference ellipsoid was not taken.

2. Astronomic observations were used to provide a frame of reference that is non-orthogonal because of the convergence of the meridians. The result is the fact that straight lines on the ground are lines with constantly changing bearings, meridians excepted.

3. The east and west lines of a township are not parallel due to the convergence of the meridians.

4. If we were to define a perfect traverse as having no random error component, a perfect traverse to layout a township would not be expected to close because of the convergence of the meridians.

To say that the USPLSS has geodetic implications is an understatement. Wahl and Hintz assert that the USPLSS should be considered as a separate datum because it represents a set of measurements and computations unique in definition and practice.

In the USPLSS states, land surveyors regularly perform boundary retracement surveys involving USPLSS monumentation. This chapter points to the fact that the understanding of geodesy is necessary to understand and use the USPLSS system in current land survey practice. We refer back to a statement in our introductory chapter: Not all geodesists are land surveyors, but all land surveyors are forced to be geodesists if they really want to understand the USPLSS!

Study Questions

1. Given that the area of 16 townships form a quadrangle. Consider two quadrangles located north of their respective baselines. Let the south boundary of the first quadrangle be located on the parallel of latitude at 30° 10' 37" N and the south boundary of the second quadrangle be located on the parallel of latitude at 45° 56' 39" N. (Note: Use the GRS 80 geometric reference surface).
   a) Compute the angular and the linear convergence experienced by each quadrangle.
   b) Compare the angular and linear convergence experienced by the two quadrangles. Compute the difference in the angular convergence and then compute the difference in the linear convergence. What conclusions can you draw from the comparison?

2. Recompute the study question above using the sphere as a reference surface. Assume a radius of 6,371,000 meters. Are the differences significant? Explain your answer.
3. Why don't you need to know the length of the meridian to compute angular convergence?
4. A United States Geological Survey 7.5' topographic map (quadrangle) covers an area of 7.5' of latitude by 7.5' of longitude. Compute the convergence experienced by two USGS Topographic Maps. Chewela is the description of a 7.5 minute map in Washington State that has the latitude of 48° 15' 00" N along its southern edge. Gulf Shores is the description of a 7.5 minute map in Alabama that has the latitude of 30° 15' 00" N along its southern edge.
   a) Compute the angular convergence and linear convergence experienced by each topographic map. Use the Clarke spheroid of 1866 as a geometric reference surface.
   b) Use your computations to answer the following questions:
      (1) Is the length of the parallel covered by the 7.5' on the southern boundary of each map the same? Why or why not?
      (2) If you wanted the same convergence angle in Chewela as you computed in Gulf Shores, would the topographic map for Chewela have to be bigger or smaller?
      (3) If you were a surveyor in the early 1800's using equipment of low precision, where would the effect of convergence most likely to be seen in the field measurements: Indiana or Alabama? Why?
5. Problems 6.1 and 6.2 in the body of the chapter computed the convergence between two meridians on the sphere and then on the ellipsoid of revolution. The difference was only one tenth of a second of angular convergence. Compute the distance in feet on the ground represented by that one tenth of a second. Is this value dependent on orthometric height?
6. If you were a legislator representing the surveying community in 1785 and you knew what you know now about the effect of convergence, what changes would you have made to the USPLSS? How would the changes you propose have changed the final outcome we live with today?
7. Pursuant to Sec. 3-48 of the 1973 BLM Manual, meridional section lines were to be run parallel with the east boundary of the township when subdividing the township. Assuming that the east township boundary had a bearing of North, what would be the bearing of the west line of section 36 if the south township boundary has a geodetic latitude of 40°N? The west boundary of section 8 for the same township? Use the GRS 80 ellipsoid.
8. Section 2-77 of the 1973 BLM Manual describes the tangent method for establishing a parallel on the earth's surface. Figure 13 in the Manual shows offsets from the tangent to the parallel for a given latitude.
   a) Verify the offset from the tangent to the southeast corner of section 35 as shown in said figure. Assume a spherical earth having a radius of 6371 km.
   b) Verify the azimuth (bearing) of the tangent three miles east of the southwest corner of the township.

CHAPTER SEVEN

GEODETIC REFERENCE SYSTEMS

Rotations preserve the length of a vector. So do reflections. So do permutations. So does multiplication by any orthogonal matrix--*lengths do not change*. This a key property of Q (a rotation matrix): If Q has orthonormal columns (QᵀQ = I), it leaves the lengths unchanged;

$$\|Qx\| = \|x\| \text{ for every vector x.}$$

Q also preserves dot products and angles: (Qx)ᵀ(Qy) = xᵀQᵀQy = xᵀy

Proof: $\|Qx\|^2$ is the same as $\|x\|^2$ because $(Qx)^T(Qx) = x^T Q^T Q x = x^T I x.$

Orthogonal matrices are excellent for computations--numbers can never grow too large when lengths are fixed. Good computer codes use Q's as much as possible, to be numerically stable.
Strang and Borre, 1997

The *reference systems* or *reference frames* used to express point positions and observations in geodesy are nothing more than coordinate systems. Like the spherical coordinate systems presented in Chapter 3, the geodetic reference systems make use of curvilinear and Cartesian (rectangular) coordinate systems that are referred to the ellipsoid. Coordinate transformations are also necessary to transform coordinates in one system to another.

Coordinate Systems

We will use three different 3-D coordinate systems as a means of expressing point positions and other relevant quantities with respect to the ellipsoid. All three systems have useful applications with each system being handier in certain situations than in others. The user must understand the differences between and relationships among the three systems.

Geodetic (Curvilinear) Coordinates

The *geodetic coordinates* of a point are expressed using the curvilinear values of geodetic longitude (λ), geodetic latitude (ϕ) and height above or below the ellipsoid surface (h_e). These values form a right-handed, earth-fixed, 3-D coordinate system where point positions may be expressed using the coordinate triplet (λ, ϕ, h_e). Geodetic longitude is reckoned in the same manner as

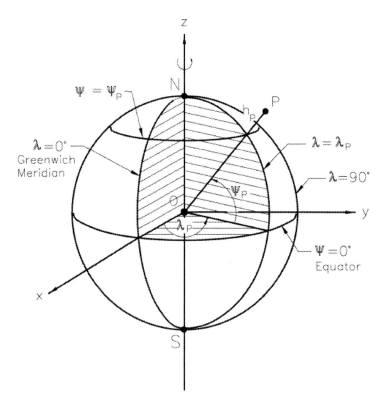

Figure 7.1 Geodetic (curvilinear) coordinates for point P.

spherical longitude and is expressed as positive to the east. Geodetic latitude is reckoned from the plane of the equator to the normal at a point and is expressed as positive to the north. Ellipsoid height (h_e) is expressed in meters. Figure 7.1 illustrates these values for a point located above the ellipsoid surface.

Geocentric (Cartesian) Coordinates

The *geocentric coordinates* of a point are expressed using rectangular Cartesian coordinates (x, y, z) referred to an earth-centered (geocentric), earth-fixed, right-handed, orthogonal, 3-D axis system. This system is sometimes referred to as the *ECEF system*. The orientation of the axes is identical to that used in the spherical earth model. The origin of the coordinate system is at the earth's *center of mass* that corresponds with the center of the ellipsoid (intersection of equatorial plane and axis of rotation). The x-axis lies in the equatorial plane with its positive end intersecting the Greenwich meridian. The y-axis lies in the equatorial plane with its positive end intersecting the ellipsoid at 90° E longitude. The z-axis is coincident with the earth's spin axis, positive toward the North Pole. Figure 7.1 depicts the location of the axes.

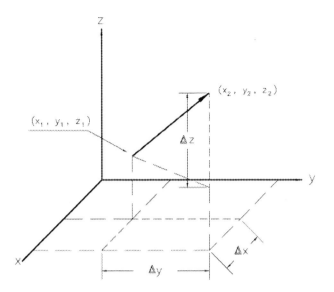

Figure 7.2 GPS vector in 3-space.

GPS vectors are described using differences in geocentric coordinates (Δx, Δy, Δz). Figure 7.2 illustrates a GPS vector in 3-D space with its components.

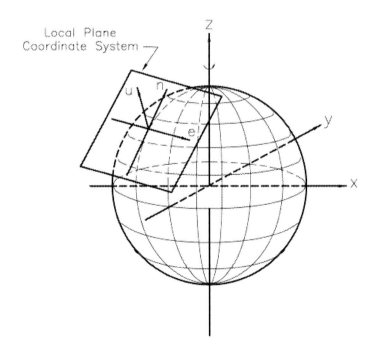

Figure 7.3 Local geodetic coordinate system.

Local Geodetic Horizon Coordinates

Local Geodetic Horizon Coordinates or *local geodetic coordinates* are extremely useful when integrating GPS determined positions with terrestrial (e.g., total station) observations. This system is sometimes referred to as the *LGH system*. The local geodetic coordinate system is an earth-fixed, right-handed, orthogonal, 3-D coordinate system having its origin at any point specified. The north axis lies in the meridian plane and is directed positive toward the North Pole. The up axis (sometimes called the *h* axis) lies

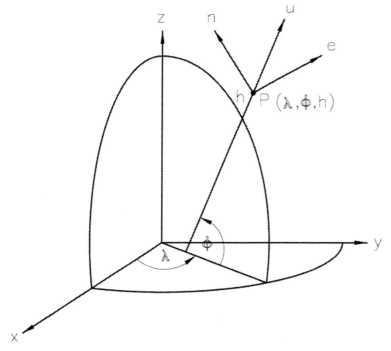

Figure 7.4 Local geodetic coordinate system (e, n, u).

along a normal to the ellipsoid at the origin, positive outside the ellipsoid surface. The east axis forms the right-handed system by being perpendicular to the meridian plane, positive to the east. Figure 7.3 shows the orientation of a local geodetic coordinate system that is tangent to the ellipsoid surface. Note how the e-n plane coincides with the local geodetic horizon. Figure 7.4 illustrates a local geodetic coordinate system that is located above the surface of the ellipsoid. A point position is expressed as an (e, n, u) triplet. Note that the origin of the system has local geodetic coordinates (0, 0, 0). A confusing aspect of this system is that an infinite number of local geodetic systems may be established, typically, at each instrument setup.

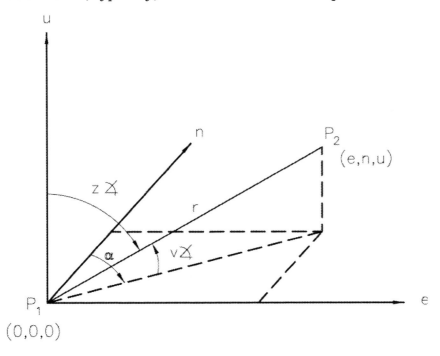

Figure 7.5 Points tied to origin using azimuth, zenith angle, and slant range.

The local geodetic coordinates of any point are referred to the origin of the local geodetic system using the *geodetic azimuth* (α), *vertical angle* ($v\angle$) or *zenith angle* ($z\angle$) and *mark-to-mark slant range* (r) from the origin to the point (figure 7.5). Right angle trigonometry **may be** used to define the following relations:

$$e = r\cos(v\angle)\sin\alpha = r\sin(z\angle)\sin\alpha$$
$$n = r\cos(v\angle)\cos\alpha = r\sin(z\angle)\cos\alpha$$
$$u = r\sin(v\angle) = r\cos(z\angle)$$

$$\alpha = \arctan\left(\frac{e}{n}\right)$$

$$r = \left(e^2 + n^2 + u^2\right)^{1/2}$$

$$v\angle = \arcsin\left(\frac{u}{r}\right)$$

$$z\angle = \arccos\left(\frac{u}{r}\right)$$

Problem 7.1: Station Low is the origin of a local geodetic horizon coordinate system. Compute the local geodetic horizon coordinates of station Echo if the following observations are made from Low to station Echo:

geodetic azimuth = 240°00'00"
zenith angle = 92°00'00"
slant range = 1000.000 m

$$e = 1000.000 \text{m} \sin 92°00'00" \sin 240°00'00" = -865.498 \text{m}$$

$$n = 1000.000 \text{m} \sin 92°00'00" \cos 240°00'00" = -499.695 \text{m}$$

$$u = 1000.000 \text{m} \cos 92°00'00" = -34.899 \text{m}$$

Coordinate Transformations

Coordinate transformations are the mathematical formulas used to transform coordinates in one reference system to another.

Geodetic to Geocentric

To convert from geodetic coordinates (λ, ϕ, h_e) to geocentric coordinates (x, y, z), use the following formulas. Remember that N is the radius of curvature in the prime vertical and don't forget that longitude is positive to the east!

$$(x, y, z) = f(\lambda, \phi, h_e)_{a,f}$$
$$x = (N + h_e) \cos \phi \cos \lambda$$
$$y = (N + h_e) \cos \phi \sin \lambda$$
$$z = [N(1 - e^2) + h_e] \sin \phi$$

Problem 7.2: Compute the geocentric coordinates for station Simple given the following GRS 80 geodetic coordinates: λ = 80°00'00.0000" W, ϕ = 40°00'00.0000" N, and h = 100.000 m.

$$N = \frac{6,378,137 \ m}{[1 - (0.0066943800) \sin^2 40°]^{\frac{1}{2}}} = 6,386,976.166 \ m$$

$$x = (6,386,976.166 \ m + 100 \ m) \cos 40° \cos(80°) = 849,623.061 \ m$$
$$y = (6,386,976.166 \ m + 100 \ m) \cos 40° \sin(80°) = -4,818,451.818 \ m$$
$$z = [6,386,976.166 \ m(1 - 0.0066943800) + 100 \ m] \sin 40° = 4,078,049.851 \ m$$

Geocentric to Geodetic

To convert from geocentric coordinates (x, y, z) to geodetic coordinates (λ, ϕ, h_e) an iterative process must be employed. The following equations are used:

$$\left(\lambda, \phi, h_e\right)_{a,f} = g(x, y, z)$$

$$h_e = \frac{\sqrt{x^2 + y^2}}{\cos\phi} - N$$

$$\phi = \arctan\left\{\frac{z}{\sqrt{x^2 + y^2}}\left[1 - e^2\left(\frac{N}{N + h_e}\right)\right]^{-1}\right\}$$

$$\lambda = \arctan\left(\frac{y}{x}\right)$$

Iteration is required because ϕ and h_e are dependent upon one another. In order to solve for these two dependent values, first assume $h_e = 0$ and solve for ϕ. Next solve for h_e and solve once again for ϕ. Continue the iteration until the change in ϕ between successive iterations is sufficiently minuscule.

Problem 7.3: Compute the GRS 80 geodetic coordinates for a station having geocentric coordinates x = -2,490,000.000 m, y = -4,020,000.000 m, and z = 4,267,000.000 m.

First iteration (subscript identifies iteration)
Assume $h_1 = 0.000$ m.

$$\phi_1 = \arctan\left\{\left[\frac{4{,}267{,}000\ m}{\sqrt{(-2{,}490{,}000\ m)^2 + (-4{,}020{,}000\ m)^2}}\right](1 - 0.0066943800)^{-1}\right\}$$

$$\phi_1 = 42°15'12.39711''$$

$$N_1 = \frac{6{,}378{,}137.\ m}{\left[1 - (0.00669438002)\sin^2 42°15'12.39711''\right]^{1/2}} = 6{,}387{,}811.565\ m$$

Second iteration

$$h_2 = \frac{\sqrt{(-2{,}490{,}000 \text{ m})^2 + (-4{,}020{,}000 \text{ m})^2}}{\cos 42°15'12.39711''} - 6{,}387{,}811.565 \text{ m} = 774.500 \text{ m}$$

$$\phi_2 = \arctan\left\{\left[\frac{4{,}267{,}000\,m}{\sqrt{(-2{,}490{,}000\,m)^2 + (-4{,}020{,}000\,m)^2}}\right](1 - 0.0066943800)\left(\frac{6{,}387{,}811.565\,m}{6{,}387{,}811.565\,m + 774.500\,m}\right)^{-1}\right\}$$

$\phi_2 = 42°15'12.31323''$

$N_2 = 6{,}387{,}811.557$ m

Third iteration

$h_3 = 772.149$ m

$\phi_3 = 42°15'12.31349''$

$N_3 = 6{,}387{,}811.557$ m

Fourth iteration

$h_4 = 772.156$ m

$\phi_4 = 42°15'12.31349''$ (stable)

$N_4 = 6{,}387{,}811.557$ m

Fifth iteration

$h_5 = 772.156$ m (stable)

$$\lambda = \arctan\left(\frac{-4{,}020{,}000 \text{ m}}{-2{,}490{,}000 \text{ m}}\right) = 58°13'32.88819'' + 180° \text{ (third quadrant)} = 238°13'32.88819''$$

$\therefore$ $\lambda = 238°13'32.88819''$ E = $121°46'27.11181''$ W

Note that the determination of longitude requires the student to evaluate the algebraic signs of both the numerator and denominator in order to add a quadrant correction for quadrants two, three and four.

Generic Coordinate Transformations

The topic of coordinate transformations doesn't often come up in conversations, except perhaps among students agonizing over a surveying assignment! Actually though, surveyors perform coordinate transformations repeatedly when analyzing survey data for boundary surveys. Coordinate transformations are also an intimate part of GPS surveying, although you may not recognize them. Coordinate transformations may also be implemented to solve a number of extremely useful surveying problems.

Two Dimensional (2-D) Coordinate Transformations

It is easiest to think in terms of two-dimensions (2-D) so we will start there. Curious, isn't it, that we live in a three-dimensional world (3-D), but are taught to think in 2-D only? Even as surveyors we like to break our job up into 2-D tasks (e.g., plane traversing) and 1-D tasks (e.g., leveling).

Consider a 2-D coordinate system. The axes may be labeled *N* and *E* or *x* and *y*. I'll use the latter since they lead into a discussion of 3-D coordinates more readily. Any point position may be expressed in the *x-y* system by simply listing the corresponding *rectangular coordinate* pair (*x*, *y*). This effectively locates the point with respect to the origin of the coordinate system - whether we know where the origin is (physically) or not. The position of the point may also be expressed in terms of *polar coordinates* (r, γ) where r is the distance from the origin to the point and γ is the angle measured with respect to the *x*-axis (positive counterclockwise). Accordingly, point P in figure 7.6 may be expressed by either rectangular or polar coordinates. The relationship between rectangular and polar coordinates may be expressed:

x = r cos γ

y = r sin γ

Rotation

Now, what happens when you pick the "rotate" function in your coordinate geometry (COGO) program? We often think or say that we are "rotating the points". This is incorrect reasoning! Do the physical points move when I hit the rotate button? I hope not! Things could get pretty messy if they did! Actually, the coordinate axes are being rotated and, therefore, new coordinate values are assigned to the points. How does this come about? Dust off your old trigonometry book and let's take a look.

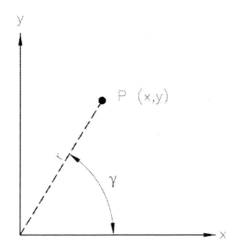

Figure 7.6 Use of polar coordinates.

Figure 7.7 shows the original *x-y* coordinate axes and the rotated *x'-y'* coordinate axes. The latter system has been rotated by an angle of θ with respect to the former system. The rotation is considered positive if the new system is rotated counterclockwise with respect to the old system. The figure therefore illustrates a positive rotation. Point P may be described in either

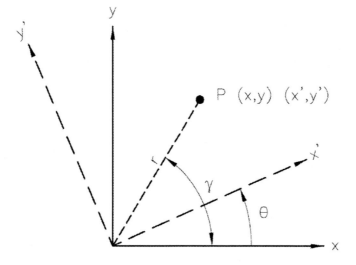

Figure 7.7 Rotation.

rectangular system by coordinates (x, y) or (x', y'). The relationship between the x'-y' system and the x-y system may be expressed in terms of the rotation angle by making use of trigonometric formulas for the difference of two angles and polar coordinates.

The rotation angle answers the question: "How must I rotate the non-prime system in order to align it with the prime system?"

$x' = r \cos (\gamma - \theta)$

$\quad = r (\cos\gamma \cos\theta + \sin\gamma \sin\theta)$

$\quad = r \cos\gamma \cos\theta + r \sin\gamma \sin\theta$

$x' = x \cos\theta + y \sin\theta$

$y' = r \sin (\gamma - \theta)$

$\quad = r (\sin\gamma \cos\theta - \cos\gamma \sin\theta)$

$\quad = r \sin\gamma \cos\theta - r \cos\gamma \sin\theta$

$y' = -x \sin\theta + y \cos\theta$

It is handier to use matrix notation for these calculations. In shorthand form we may write $\mathbf{X'} = \mathbf{RX}$ where $\mathbf{X'}$ is a column vector of rotated coordinates, $\mathbf{R}$ is the rotation matrix, and $\mathbf{X}$ is a column vector of pre-rotated coordinates. This may be written in matrix form as:

$$\begin{bmatrix} x' \\ y' \end{bmatrix} = \begin{bmatrix} \cos\theta & \sin\theta \\ -\sin\theta & \cos\theta \end{bmatrix} \begin{bmatrix} x \\ y \end{bmatrix}$$

or even simplified further by *reparameterization* as:

$$\begin{bmatrix} x' \\ y' \end{bmatrix} = \begin{bmatrix} a & b \\ -b & a \end{bmatrix} \begin{bmatrix} x \\ y \end{bmatrix}$$

Problem 7.4: A point has (x, y) coordinates of $(-5, 3)$. The coordinate axes are then rotated 135° clockwise. Determine the rotated coordinates of the point.

The clockwise rotation represents a rotation angle of -135° or +225°. This rotation angle is inserted into the matrix equation, along with the original coordinates to compute the rotated coordinates of the point.

$$\begin{bmatrix} x' \\ y' \end{bmatrix} = \begin{bmatrix} \cos 225° & \sin 225° \\ -\sin 225° & \cos 225° \end{bmatrix} \begin{bmatrix} -5 \\ 3 \end{bmatrix} = \begin{bmatrix} 1.4 \\ -5.7 \end{bmatrix}$$

If you are having trouble visualizing this problem, draw the problem to scale showing both the original coordinate axes and the rotated coordinate axes. Scale off the rotated coordinates as a check!

Translation

The translation function performed by your COGO program is even a simpler function. Figure 7.8 shows our original x-y system again, just like in

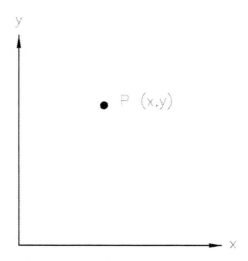

Figure 7.8 Coordinate axes before translation.

figure 7.6. When we translate the coordinate axes (not the points!), we end up with a scenario as shown in figure 7.9. The x'-y' system represents the translated system. Two translations have been performed, one in the x direction and one in the y direction. The translations are abbreviated t_x and t_y, respectively. The algebraic sign of the translations is dependent on whether the translations are expressed in the x-y system (positive) or the x'-y' system (negative). The x'-y' system coordinates are derived simply by subtracting the translations from their corresponding x-y system coordinates:

$x' = x - t_x$

$y' = y - t_y$

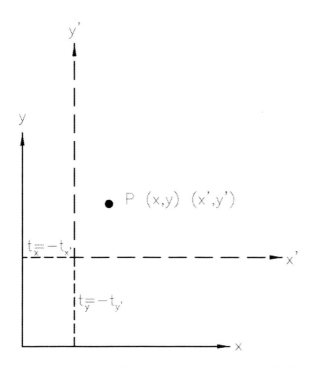

Figure 7.9 Coordinate axes after translation.

$(t_x$ & t_y in x-y system) or in matrix shorthand $\mathbf{X'} = \mathbf{X} - \mathbf{T}$ where $\mathbf{T}$ is the translation vector expressed in the pre-translation system. In traditional matrix form the two equations may be written:

$$\begin{bmatrix} x' \\ y' \end{bmatrix} = \begin{bmatrix} x \\ y \end{bmatrix} - \begin{bmatrix} t_x \\ t_y \end{bmatrix}$$

Alternatively, the equations may be written:

$$x' = x + t_{x'}$$
$$y' = y + t_y$$

$(t_{x'}$ & $t_{y'}$ in x'-y' system) or,

$$\begin{bmatrix} x' \\ y' \end{bmatrix} = \begin{bmatrix} x \\ y \end{bmatrix} + \begin{bmatrix} t_{x'} \\ t_{y'} \end{bmatrix}$$

109

which may be written $\mathbf{X'} = \mathbf{X} + \mathbf{T'}$ where $\mathbf{T'}$ is the translation vector expressed in the translated system. We'll stick with the use of this latter form since it will be more convenient to use it a bit later.

Problem 7.5: A coordinate system is translated 5 units in the x direction and −12 units in the y direction relative to its original location. Determine the translated coordinates of a point having original coordinates (6,8).

The tricky part of translations is to determine the coordinate system in which the translation is being expressed. By the problem statement we know that the translation is expressed in the original system so $t_x = 5$ and $t_y = -12$. Therefore, we compute the translated coordinates using the form containing t_x and t_y.

$$\begin{bmatrix} x' \\ y' \end{bmatrix} = \begin{bmatrix} x \\ y \end{bmatrix} - \begin{bmatrix} t_x \\ t_y \end{bmatrix} = \begin{bmatrix} 6 \\ 8 \end{bmatrix} - \begin{bmatrix} 5 \\ -12 \end{bmatrix} = \begin{bmatrix} 1 \\ 20 \end{bmatrix}$$

Since no rotation was involved in this problem, the translation could easily be expressed in the translated (prime) system by changing the signs of the previously used translation values. Please be aware that this is only true when no rotations are involved.

$$\begin{bmatrix} x' \\ y' \end{bmatrix} = \begin{bmatrix} x \\ y \end{bmatrix} + \begin{bmatrix} t_{x'} \\ t_{y'} \end{bmatrix} = \begin{bmatrix} 6 \\ 8 \end{bmatrix} + \begin{bmatrix} -5 \\ 12 \end{bmatrix} = \begin{bmatrix} 1 \\ 20 \end{bmatrix}$$

Scale Change

The other type of coordinate transformation you may wish to perform is a scale change. This may come about if you need to convert from feet to meters,

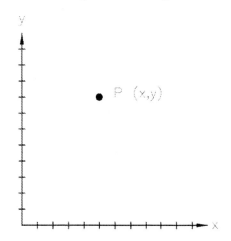

Figure 7.10 Coordinate axes before change in scale.

or vice versa. Alternatively, you may think of this as the factor used to "adjust your chain" to another surveyor's measurements in retracement work. Regardless, it involves differing graduations along the original *x*-*y* system axes and the transformed *x'*-*y'* system axes.

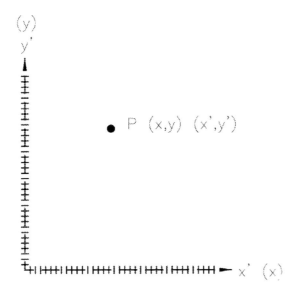

Figure 7.11 Coordinate axes after adjustment for scale.

This is illustrated in figures 7.10 and 7.11. The relationship between the two systems may be expressed as:

$$x' = s\, x$$

$$y' = s\, y$$

where s is a scale factor. These equations may also be written as:

$$\mathbf{X'} = \mathbf{sX} \quad \text{or} \quad \begin{bmatrix} x' \\ y' \end{bmatrix} = s \begin{bmatrix} x \\ y \end{bmatrix}$$

Problem 7.6: A point has coordinates (1.000 m, 1.829 m). Compute the point's coordinates if the units are changed to international feet.

$$\begin{bmatrix} x' \\ y' \end{bmatrix} = s \begin{bmatrix} x \\ y \end{bmatrix} = \left(\frac{1\,\text{int'l ft}}{0.3048\,\text{m}} \right) \begin{bmatrix} 1.000\,\text{m} \\ 1.829\,\text{m} \end{bmatrix} = \begin{bmatrix} 3.28\,\text{int'l ft} \\ 6.00\,\text{int'l ft} \end{bmatrix}$$

111

Four Parameter Transformation

These three individual 2-D coordinate transformations may be combined into two equations that are known as a *two-dimensional conformal coordinate transformation* or a *four-parameter transformation*:

$$x' = s \ (x \ cos\theta + y \ sin\theta) + t_{x'}$$

$$y' = s \ (-x \ sin\theta + y \ cos\theta) + t_{y'}$$

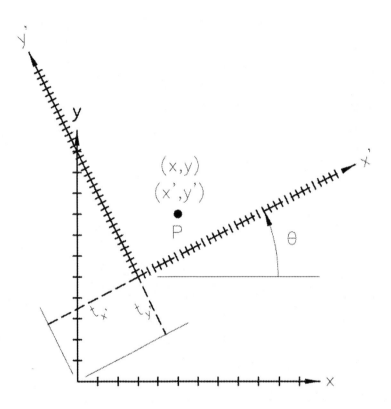

Figure 7.12 Four-Parameter Transformation.

As before, these equations may be written as $\mathbf{X'} = s\mathbf{RX} + \mathbf{T'}$ or $\mathbf{X'} = s\mathbf{R(X-T)}$ in shorthand notation or using standard notation:

$$\begin{bmatrix} x' \\ y' \end{bmatrix} = s \begin{bmatrix} cos\theta & sin\theta \\ -sin\theta & cos\theta \end{bmatrix} \begin{bmatrix} x \\ y \end{bmatrix} + \begin{bmatrix} t_{x'} \\ t_{y'} \end{bmatrix} \quad or \quad \begin{bmatrix} x' \\ y' \end{bmatrix} = s \left\{ \begin{bmatrix} cos\theta & sin\theta \\ -sin\theta & cos\theta \end{bmatrix} \left(\begin{bmatrix} x \\ y \end{bmatrix} - \begin{bmatrix} t_x \\ t_y \end{bmatrix} \right) \right\}$$

or we may reparameterize once again and come up with:

$$\begin{bmatrix} x' \\ y' \end{bmatrix} = \begin{bmatrix} a & b \\ -b & a \end{bmatrix} \begin{bmatrix} x \\ y \end{bmatrix} + \begin{bmatrix} c \\ d \end{bmatrix}$$

Figure 7.12 shows the general four-parameter transformation. Note that the translations are shown in the x'-y' system. A conformal transformation is one in which the coordinate axes remain perpendicular to one another. The four parameters refer to the rotation angle, θ, two translations, $t_{x'}$ & $t_{y'}$, and the scale factor, s. These two equations allow the surveyor to compute the coordinates of points in the transformed (rotated, translated and/or scaled) x'-y' coordinate system given the coordinates of the points in the original x-y system and the four transformation parameters. It is unlikely that your coordinate geometry program has a coordinate transformation routine that incorporates these four parameters in a single transformation. It could easily be accomplished but it may be more confusing than valuable!

Problem 7.7: Determine the transformed (prime) coordinates for points A, B, C and D given the scenario shown below. Assume that the units are consistent in both the original and transformed coordinate systems (i.e., s = 1).

Point	x	y	x'	y'
A	4	8		
B	0	0		
C	12	12		
D	-8	6		

The first step is to determine the translation since that will determine the form of the transformation equation to use. The translation vector to be used is one that doesn't need to be calculated –- it's just easier that way! The translation vector is simply the coordinates of one origin, expressed in the other system. Looking at the figure above, the origin of the transformed system (point A) has coordinates (4, 8) in the original system, thus, $t_x = 4$ and $t_y = 8$. This would be the translation vector to use. It is possible to express the origin of the original system (point B) in the transformed system, but it would involve some trigonometry. It is easier to use the clearly expressed translation vector. The chosen translation vector requires that we use the transformation form:

$$\begin{bmatrix} x' \\ y' \end{bmatrix} = s \begin{bmatrix} \cos\theta & \sin\theta \\ -\sin\theta & \cos\theta \end{bmatrix} \left\{ \begin{bmatrix} x \\ y \end{bmatrix} - \begin{bmatrix} t_x \\ t_y \end{bmatrix} \right\}$$

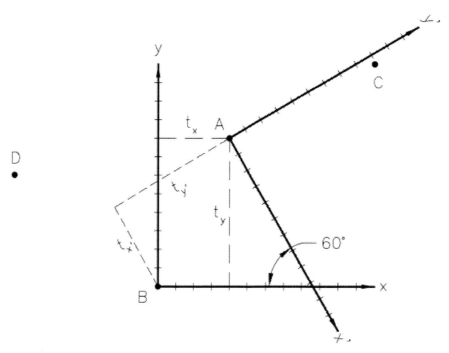

Figure 7.13 Sketch for Problem 7.7.

We still must determine the rotation angle. The magnitude of the rotation angle is 60° as is apparent in the figure. The algebraic sign of the rotation can be confusing. In order to align the original system with the transformed system, we would have to rotate the original system in a clockwise direction. Thus, the rotation is -60°. Once the translations, rotation and scale factor have been determined, fill in the transformation form with these constants, plug in the original coordinates, one-at-a-time, and compute the transformed coordinates.

$$\begin{bmatrix} x' \\ y' \end{bmatrix} = 1 \begin{bmatrix} \cos(-60°) & \sin(-60°) \\ -\sin(-60°) & \cos(-60°) \end{bmatrix} \left\{ \begin{bmatrix} x \\ y \end{bmatrix} - \begin{bmatrix} 4 \\ 8 \end{bmatrix} \right\} = \begin{bmatrix} 0.500 & -0.866 \\ 0.866 & 0.500 \end{bmatrix} \left\{ \begin{bmatrix} x \\ y \end{bmatrix} - \begin{bmatrix} 4 \\ 8 \end{bmatrix} \right\}$$

Point A
$$\begin{bmatrix} x' \\ y' \end{bmatrix} = \begin{bmatrix} 0.500 & -0.866 \\ 0.866 & 0.500 \end{bmatrix} \left\{ \begin{bmatrix} 4 \\ 8 \end{bmatrix} - \begin{bmatrix} 4 \\ 8 \end{bmatrix} \right\} = \begin{bmatrix} 0.500 & -0.866 \\ 0.866 & 0.500 \end{bmatrix} \begin{bmatrix} 0 \\ 0 \end{bmatrix} = \begin{bmatrix} 0 \\ 0 \end{bmatrix}$$

This makes sense because point A is the origin of the transformed system.

Point B
$$\begin{bmatrix} x' \\ y' \end{bmatrix} = \begin{bmatrix} 0.500 & -0.866 \\ 0.866 & 0.500 \end{bmatrix} \left\{ \begin{bmatrix} 0 \\ 0 \end{bmatrix} - \begin{bmatrix} 4 \\ 8 \end{bmatrix} \right\} = \begin{bmatrix} 0.500 & -0.866 \\ 0.866 & 0.500 \end{bmatrix} \begin{bmatrix} -4 \\ -8 \end{bmatrix} = \begin{bmatrix} 4.93 \\ -7.46 \end{bmatrix}$$

114

The coordinates of point B, being the origin of the original system expressed in the transformed system, are $t_{x'}$ and $t_{y'}$.

Point C

$$\begin{bmatrix} x' \\ y' \end{bmatrix} = \begin{bmatrix} 0.500 & -0.866 \\ 0.866 & 0.500 \end{bmatrix} \left\{ \begin{bmatrix} 12 \\ 12 \end{bmatrix} - \begin{bmatrix} 4 \\ 8 \end{bmatrix} \right\} = \begin{bmatrix} 0.500 & -0.866 \\ 0.866 & 0.500 \end{bmatrix} \begin{bmatrix} 8 \\ 4 \end{bmatrix} = \begin{bmatrix} 0.54 \\ 8.93 \end{bmatrix}$$

Point D

$$\begin{bmatrix} x' \\ y' \end{bmatrix} = \begin{bmatrix} 0.500 & -0.866 \\ 0.866 & 0.500 \end{bmatrix} \left\{ \begin{bmatrix} -8 \\ 6 \end{bmatrix} - \begin{bmatrix} 4 \\ 8 \end{bmatrix} \right\} = \begin{bmatrix} 0.500 & -0.866 \\ 0.866 & 0.500 \end{bmatrix} \begin{bmatrix} -12 \\ -2 \end{bmatrix} = \begin{bmatrix} -4.27 \\ -11.4 \end{bmatrix}$$

The transformed coordinates for points C and D seem reasonable based upon the figure.

Three Dimensional (3-D) Coordinate Transformations

The 2-D coordinate transformations just discussed consist of four parameters: 1 rotation, 2 translations, and 1 scale change. Since 3-D coordinate systems have three axes, three rotations are possible; translations may be made along any of the coordinate axes, therefore three translations are possible. To make matters simple, we will consider that any scale change affects all axes the same. Thus, we will only consider three-dimensional coordinate transformations that consist of these seven parameters: 3 rotations, 3 translations and 1 scale change.

The standard format for the *seven-parameter transformation (3-D conformal coordinate transformation)* is identical in shorthand notation to the four-parameter transformation $\mathbf{X'} = s\mathbf{R}\mathbf{X} + \mathbf{T'}$ but differences do show up in the standard matrix form:

$$\begin{bmatrix} x' \\ y' \\ z' \end{bmatrix} = s\mathbf{R} \begin{bmatrix} x \\ y \\ z \end{bmatrix} + \begin{bmatrix} t_{x'} \\ t_{y'} \\ t_{z'} \end{bmatrix}$$

Rotation

The rotation matrix denoted by $\mathbf{R}$ in the previous equation requires detailed explanation. The 3-D rotation matrix is the product of two or more single axis rotation matrices. A *single axis rotation matrix* is the rotation matrix that describes the effect of rotating the entire coordinate system about an axis through a specified angle. All rotation matrices used herein are square, *normal orthogonal* matrices. The transpose of a normal orthogonal matrix is equivalent to its inverse. The general forms for the single axis rotation matrices are:

Rotation About First (x) Axis

A rotation about the first axis through an angle α produces the first axis rotation matrix $\mathbf{R_1}$ having the standard form:

$$\mathbf{R}_1 = \begin{bmatrix} 1 & 0 & 0 \\ 0 & \cos\alpha & \sin\alpha \\ 0 & -\sin\alpha & \cos\alpha \end{bmatrix}$$

Problem 8: Prove that the first axis coordinates are preserved by a first axis rotation. This will be shown by the use of a simple 3-D coordinate transformation.

$$\begin{bmatrix} x' \\ y' \\ z' \end{bmatrix} = \begin{bmatrix} 1 & 0 & 0 \\ 0 & \cos\alpha & \sin\alpha \\ 0 & -\sin\alpha & \cos\alpha \end{bmatrix} \begin{bmatrix} x \\ y \\ z \end{bmatrix} = \begin{bmatrix} x \\ y\cos\alpha + z\sin\alpha \\ z\cos\alpha - y\sin\alpha \end{bmatrix} \qquad \therefore x' = x$$

Rotation About Second (y) Axis

A rotation about the second axis through an angle β produces the second axis rotation matrix $\mathbf{R_2}$ having the standard form:

$$\mathbf{R}_2 = \begin{bmatrix} \cos\beta & 0 & -\sin\beta \\ 0 & 1 & 0 \\ \sin\beta & 0 & \cos\beta \end{bmatrix}$$

Rotation About Third (z) Axis

A rotation about the third axis through an angle γ produces the third axis rotation matrix $\mathbf{R_3}$ having the standard form:

$$\mathbf{R}_3 = \begin{bmatrix} \cos\gamma & \sin\gamma & 0 \\ -\sin\gamma & \cos\gamma & 0 \\ 0 & 0 & 1 \end{bmatrix}$$

As stated $\mathbf{R} = f\{\mathbf{R}_1, \mathbf{R}_2, \mathbf{R}_3\}$ where f is the matrix product. The order of the single axis rotations is critical! The first single axis rotation matrix is *premultiplied* by the second single axis rotation matrix. Successive single axis rotation matrices are *postmultiplied* by preceding single axis rotation matrices, i.e., the single axis rotation matrices are written right to left, from first to last. For example, assume a 3-D transformation consists of four rotations - the first about the y-axis through an angle of 35°, the second about the x-axis through an angle of -135°, the third about the y-axis of 85°, and the fourth about the z-axis through an angle of 15°. The rotation matrix formed by these single axis rotations would be $\mathbf{R} = \mathbf{R}_3(15°)\mathbf{R}_2(85°)\mathbf{R}_1(-135°)\mathbf{R}_2(35°)$.

The algebraic sign of the rotation angle is considered positive if the rotation is viewed as a counterclockwise rotation from the positive end of the axis while looking toward the origin. Any rotation may be converted from negative to positive by adding 360° to the rotation angle.

Problem 7.8: A coordinate system is rotated in the following amounts and directions (when looking at the origin from the positive end of the axis specified). Compute and simplify the rotation matrix, **R**.

1st rotation: 30° clockwise about the second axis

2nd rotation: 135° counterclockwise about the first axis

3rd rotation: 60° counterclockwise about the third axis

$\mathbf{R} = \mathbf{R}_3(60°)\,\mathbf{R}_1(135°)\,\mathbf{R}_2(-30°)$

$$\mathbf{R} = \begin{bmatrix} \cos 60° & \sin 60° & 0 \\ -\sin 60° & \cos 60° & 0 \\ 0 & 0 & 1 \end{bmatrix} \begin{bmatrix} 1 & 0 & 0 \\ 0 & \cos 135° & \sin 135° \\ 0 & -\sin 135° & \cos 135° \end{bmatrix} \begin{bmatrix} \cos(-30°) & 0 & -\sin(-30°) \\ 0 & 1 & 0 \\ \sin(-30°) & 0 & \cos(-30°) \end{bmatrix}$$

$$R = \begin{bmatrix} \cos 60° & \sin 60° \cos 135° & \sin 60° \sin 135° \\ -\sin 60° & \cos 60° \cos 135° & \cos 60° \sin 135° \\ 0 & -\sin 135° & \cos 135° \end{bmatrix} \begin{bmatrix} \cos(-30°) & 0 & -\sin(-30°) \\ 0 & 1 & 0 \\ \sin(-30°) & 0 & \cos(-30°) \end{bmatrix}$$

$r11 = \cos 60° \cos(-30°) + \sin 60° \sin 135° \sin(-30°) = 0.1268$

$r12 = \sin 60° \cos 135° = -0.6124$

$r13 = \sin 60° \sin 135° \cos(-30°) - \cos 60° \sin(-30°) = 0.7803$

$r21 = \cos 60° \sin 135° \sin(-30°) - \sin 60° \cos(-30°) = -0.9268$

$r22 = \cos 60° \cos 135° = -0.3536$

$r23 = \sin 60° \sin(-30°) + \cos 60° \sin 135° \cos(-30°) = -0.1268$

$r31 = \cos 135° \sin(-30°) = 0.3536$

$r32 = -\sin 135° = -0.7071$

$r33 = \cos 135° \cos(-30°) = -0.6124$

$$\therefore R = \begin{bmatrix} r11 & r12 & r13 \\ r21 & r22 & r23 \\ r31 & r32 & r33 \end{bmatrix} = \begin{bmatrix} 0.1268 & -0.6124 & 0.7803 \\ -0.9268 & -0.3536 & -0.1268 \\ 0.3536 & -0.7071 & -0.6124 \end{bmatrix}$$

Translation

As in 2-D translations, 3-D translations may be expressed in the original or transformed systems. Usually the latter method is most convenient, but the former may be pressed into service at times. Accordingly, simple translations may be expressed in matrix notation as:

$$\begin{bmatrix} x' \\ y' \\ z' \end{bmatrix} = \begin{bmatrix} x \\ y \\ z \end{bmatrix} + \begin{bmatrix} t_{x'} \\ t_{y'} \\ t_{z'} \end{bmatrix} \quad \text{or} \quad \begin{bmatrix} x' \\ y' \\ z' \end{bmatrix} = \begin{bmatrix} x \\ y \\ z \end{bmatrix} - \begin{bmatrix} t_x \\ t_y \\ t_z \end{bmatrix}$$

The prime system translation vector may be thought of as the origin of the "non-prime" system expressed in the prime system. Likewise, the non-prime system translation vector is the origin of the prime system expressed in the non-prime system. It is important to grasp this concept!

General Form

The general form for 3-D conformal coordinate transformations may be expressed as:

$$\begin{bmatrix} x' \\ y' \\ z' \end{bmatrix} = s\mathbf{R}\begin{bmatrix} x \\ y \\ z \end{bmatrix} + \begin{bmatrix} t_{x'} \\ t_{y'} \\ t_{z'} \end{bmatrix} \qquad \text{or} \qquad \begin{bmatrix} x' \\ y' \\ z' \end{bmatrix} = s\mathbf{R}\left\{ \begin{bmatrix} x \\ y \\ z \end{bmatrix} - \begin{bmatrix} t_x \\ t_y \\ t_z \end{bmatrix} \right\}$$

Problem 7.9: The origin of the original coordinate system described in Problem 7.8 has coordinates (-163.652 m, 572.982 m, -851.904 m) in the transformed system. Compute transformed coordinates for the following points given their original system coordinates.

Point	x (US ft)	y (US ft)	z (US ft)	x' (m)	y' (m)	z' (m)
E	0.000	0.000	0.000			
F	100.000	-200.000	300.000			

Since the translation vector is expressed in the transformed system, use the form

$$\begin{bmatrix} x' \\ y' \\ z' \end{bmatrix} = s\mathbf{R}\begin{bmatrix} x \\ y \\ z \end{bmatrix} + \begin{bmatrix} t_{x'} \\ t_{y'} \\ t_{z'} \end{bmatrix} = \left(\frac{12.00\,\text{in}/\text{USft}}{39.37\,\text{in}/\text{m}}\right)\begin{bmatrix} 0.1268 & -0.6124 & 0.7803 \\ -0.9268 & -0.3536 & -0.1268 \\ 0.3536 & -0.7071 & -0.6124 \end{bmatrix}\begin{bmatrix} x \\ y \\ z \end{bmatrix} + \begin{bmatrix} -163.652\,\text{m} \\ 572.982\,\text{m} \\ -851.904\,\text{m} \end{bmatrix}$$

Point E

$$\begin{bmatrix} x' \\ y' \\ z' \end{bmatrix} = \left(\frac{12.00\,\text{in}/\text{USft}}{39.37\,\text{in}/\text{m}}\right)\begin{bmatrix} 0.1268 & -0.6124 & 0.7803 \\ -0.9268 & -0.3536 & -0.1268 \\ 0.3536 & -0.7071 & -0.6124 \end{bmatrix}\begin{bmatrix} 0.000\,\text{USft} \\ 0.000\,\text{USft} \\ 0.000\,\text{USft} \end{bmatrix} + \begin{bmatrix} -163.652\,\text{m} \\ 572.982\,\text{m} \\ -851.904\,\text{m} \end{bmatrix}$$

$$\begin{bmatrix} x' \\ y' \\ z' \end{bmatrix} = \begin{bmatrix} 0.000\,\text{m} \\ 0.000\,\text{m} \\ 0.000\,\text{m} \end{bmatrix} + \begin{bmatrix} -163.652\,\text{m} \\ 572.982\,\text{m} \\ -851.904\,\text{m} \end{bmatrix} = \begin{bmatrix} -163.652\,\text{m} \\ 572.982\,\text{m} \\ -851.904\,\text{m} \end{bmatrix}$$

The transformed coordinates for point E are as given in the problem statement.

Point F

$$\begin{bmatrix} x' \\ y' \\ z' \end{bmatrix} = \left(\frac{12.00\,\text{in}/\text{USft}}{39.37\,\text{in}/\text{m}}\right)\begin{bmatrix} 0.1268 & -0.6124 & 0.7803 \\ -0.9268 & -0.3536 & -0.1268 \\ 0.3536 & -0.7071 & -0.6124 \end{bmatrix}\begin{bmatrix} 100.000\,\text{USft} \\ -200.000\,\text{USft} \\ 300.000\,\text{USft} \end{bmatrix} + \begin{bmatrix} -163.652\,\text{m} \\ 572.982\,\text{m} \\ -851.904\,\text{m} \end{bmatrix}$$

$$\begin{bmatrix} x' \\ y' \\ z' \end{bmatrix} = \begin{bmatrix} 112.550\,\text{m} \\ -18.293\,\text{m} \\ -2.114\,\text{m} \end{bmatrix} + \begin{bmatrix} -163.652\,\text{m} \\ 572.982\,\text{m} \\ -851.904\,\text{m} \end{bmatrix} = \begin{bmatrix} -51.102\,\text{m} \\ 554.689\,\text{m} \\ -854.018\,\text{m} \end{bmatrix}$$

Local to Geocentric

The transformation of local geodetic coordinates (e, n, u) into geocentric Cartesian coordinates (x, y, z) is accomplished by employing the 3-D coordinate transformation. It should be noted that one cannot transform directly from local geodetic coordinates to geodetic coordinates (λ, ϕ, h), geocentric coordinates must be used.

Two rotations must be performed to align the local system with the geocentric system. The goal of the rotations is to make the corresponding axes of both systems parallel, e.g. first axis (e axis) of local system parallel with first axis (x axis) of geocentric system. The first rotation is illustrated in figure 7.14. This is a view perpendicular to the meridian plane. This rotation will make the u axis of the local system parallel with the z-axis of the geocentric system. This can be accomplished by a clockwise rotation about the positive e axis of

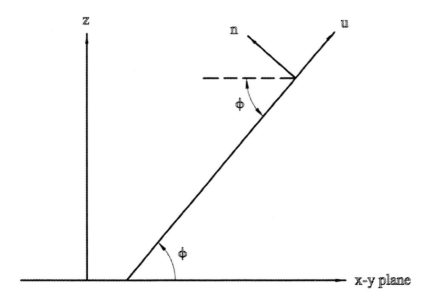

Figure 7.14 Local to geocentric transformation.

(90° - ϕ). This is a negative rotation of -(90° - ϕ) about the first axis, or simplifying, R_1(ϕ - 90°).

The second rotation will align all corresponding axes parallel with one another. Figure 7.15 is a view perpendicular to the equatorial plane. A rotation of -(90° + λ) about the u axis will accomplish this task. This rotation may be expressed as R_3(-90° - λ) or R_3(270° - λ).

The translation vector is fairly simple. The geocentric system is the prime system while the local system is the non-prime system. The origin of the non-prime (local) system expressed in the prime (geocentric) system are simply the geocentric coordinates of the origin. The transformation may be written:

$$\begin{bmatrix} x \\ y \\ z \end{bmatrix} = s\mathbf{R}_3(270° - \lambda)\mathbf{R}_1(\phi - 90°)\begin{bmatrix} e \\ n \\ u \end{bmatrix} + \begin{bmatrix} x \\ y \\ z \end{bmatrix}_{local\ origin}$$

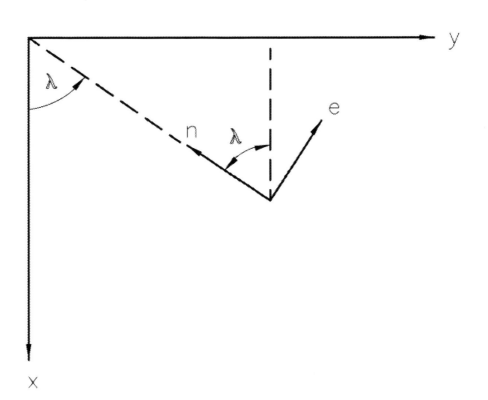

Figure 7.15 Local to Geocentric--Looking down at equatorial plane.

The transformation may be simplified further:

$$
\begin{bmatrix} x \\ y \\ z \end{bmatrix} = s \begin{bmatrix} -\sin\lambda & -\sin\phi\cos\lambda & \cos\phi\cos\lambda \\ \cos\lambda & -\sin\phi\sin\lambda & \cos\phi\sin\lambda \\ 0 & \cos\phi & \sin\phi \end{bmatrix} \begin{bmatrix} e \\ n \\ u \end{bmatrix} + \begin{bmatrix} x \\ y \\ z \end{bmatrix}_{local\ origin}
$$

Problem 7.10: The origin of a local geodetic horizon coordinate system is at station Simple (see Problem 7.2). Compute the geocentric coordinates of the following two points having local geodetic coordinates.

Point	e (m)	n (m)	u (m)	x (m)	y (m)	z (m)
G	0.000	0.000	0.000			
H	-126.348	839.609	-82.598			

See Problem 7.2 for the geodetic and geocentric coordinates for station Simple. These values have been added to the transformation shown immediately above to yield

$$
\begin{bmatrix} x \\ y \\ z \end{bmatrix} = (1) \begin{bmatrix} -\sin(-80°) & -\sin(40°)\cos(-80°) & \cos(40°)\cos(-80°) \\ \cos(-80°) & -\sin(40°)\sin(-80°) & \cos(40°)\sin(-80°) \\ 0 & \cos(40°) & \sin(40°) \end{bmatrix} \begin{bmatrix} e \\ n \\ u \end{bmatrix} + \begin{bmatrix} 849{,}623.061\,m \\ -4{,}818{,}451.818\,m \\ 4{,}078{,}049.851\,m \end{bmatrix}
$$

$$
\begin{bmatrix} x \\ y \\ z \end{bmatrix} = (1) \begin{bmatrix} 0.984808 & -0.111619 & 0.133022 \\ 0.173648 & 0.633022 & -0.754407 \\ 0 & 0.766044 & 0.642788 \end{bmatrix} \begin{bmatrix} e \\ n \\ u \end{bmatrix} + \begin{bmatrix} 849{,}623.061\,m \\ -4{,}818{,}451.818\,m \\ 4{,}078{,}049.851\,m \end{bmatrix}
$$

Point G

$$
\begin{bmatrix} x \\ y \\ z \end{bmatrix} = (1) \begin{bmatrix} 0.984808 & -0.111619 & 0.133022 \\ 0.173648 & 0.633022 & -0.754407 \\ 0 & 0.766044 & 0.642788 \end{bmatrix} \begin{bmatrix} 0.000\,m \\ 0.000\,m \\ 0.000\,m \end{bmatrix} + \begin{bmatrix} 849{,}623.061\,m \\ -4{,}818{,}451.818\,m \\ 4{,}078{,}049.851\,m \end{bmatrix}
$$

$$
\begin{bmatrix} x \\ y \\ z \end{bmatrix} = \begin{bmatrix} 0.000\,m \\ 0.000\,m \\ 0.000\,m \end{bmatrix} + \begin{bmatrix} 849{,}623.061\,m \\ -4{,}818{,}451.818\,m \\ 4{,}078{,}049.851\,m \end{bmatrix} = \begin{bmatrix} 849{,}623.061\,m \\ -4{,}818{,}451.818\,m \\ 4{,}078{,}049.851\,m \end{bmatrix}
$$

This makes sense since point G (station Simple) is the origin of the local geodetic system.

Point H

$$\begin{bmatrix} x \\ y \\ z \end{bmatrix} = (1) \begin{bmatrix} 0.984808 & -0.111619 & 0.133022 \\ 0.173648 & 0.633022 & -0.754407 \\ 0 & 0.766044 & 0.642788 \end{bmatrix} \begin{bmatrix} -126.348\,m \\ 839.609\,m \\ -82.598\,m \end{bmatrix} + \begin{bmatrix} 849{,}623.061\,m \\ -4{,}818{,}451.818\,m \\ 4{,}078{,}049.851\,m \end{bmatrix}$$

$$\begin{bmatrix} x \\ y \\ z \end{bmatrix} = \begin{bmatrix} -229.132\,m \\ 571.864\,m \\ 50.085\,m \end{bmatrix} + \begin{bmatrix} 849{,}623.061\,m \\ -4{,}818{,}451.818\,m \\ 4{,}078{,}049.851\,m \end{bmatrix} = \begin{bmatrix} 849{,}393.929\,m \\ -4{,}817{,}879.954\,m \\ 4{,}078{,}639.936\,m \end{bmatrix}$$

Geocentric to Local

The transformation from geocentric coordinates to local coordinates may be accomplished by solving the previous equation for the vector of local coordinates (*e, n, u*). This is simplified by the normal orthogonal property of the rotation matrix.

$$\begin{bmatrix} e \\ n \\ u \end{bmatrix} = \frac{1}{s} \begin{bmatrix} -\sin\lambda & \cos\lambda & 0 \\ -\sin\phi\cos\lambda & -\sin\phi\sin\lambda & \cos\phi \\ \cos\phi\cos\lambda & \cos\phi\sin\lambda & \sin\phi \end{bmatrix} \left\{ \begin{bmatrix} x \\ y \\ z \end{bmatrix} - \begin{bmatrix} x \\ y \\ z \end{bmatrix}_{local\ origin} \right\}$$

Problem 7.11: Using the same local geodetic horizon coordinate origin as in Problem 7.10 (station Simple), compute the LGH coordinates for the following point

Point	x (m)	y (m)	z (m)	e (m)	n (m)	u (m)
I	849,500.000	- 4,818,000.000	4,078,000.000			

$$\begin{bmatrix} e \\ n \\ u \end{bmatrix} = (1) \begin{bmatrix} -\sin(-80°) & \cos(-80°) & 0 \\ -\sin(40°)\cos(-80°) & -\sin(40°)\sin(-80°) & \cos(40°) \\ \cos(40°)\cos(-80°) & \cos(40°)\sin(-80°) & \sin(40°) \end{bmatrix} *$$

$$\left\{ \begin{bmatrix} 849{,}500.000\,m \\ -4{,}818{,}000.000\,m \\ 4{,}078{,}000.000\,m \end{bmatrix} - \begin{bmatrix} 849{,}623.061\,m \\ -4{,}818{,}451.818\,m \\ 4{,}078{,}049.851\,m \end{bmatrix} \right\}$$

$$\begin{bmatrix} e \\ n \\ u \end{bmatrix} = (1)\begin{bmatrix} 0.984808 & 0.173648 & 0 \\ -0.111619 & 0.633022 & 0.766044 \\ 0.133022 & -0.754407 & 0.642788 \end{bmatrix} \left\{ \begin{bmatrix} 849{,}500.000\,\text{m} \\ -4{,}818{,}000.000\,\text{m} \\ 4{,}078{,}000.000\,\text{m} \end{bmatrix} - \begin{bmatrix} 849{,}623.061\,\text{m} \\ -4{,}818{,}451.818\,\text{m} \\ 4{,}078{,}049.851\,\text{m} \end{bmatrix} \right\}$$

$$\begin{bmatrix} e \\ n \\ u \end{bmatrix} = (1)\begin{bmatrix} 0.984808 & 0.173648 & 0 \\ -0.111619 & 0.633022 & 0.766044 \\ 0.133022 & -0.754407 & 0.642788 \end{bmatrix} \begin{bmatrix} -123.061\,\text{m} \\ 451.818\,\text{m} \\ -49.851\,\text{m} \end{bmatrix} = \begin{bmatrix} -42.734\,\text{m} \\ 261.559\,\text{m} \\ -389.268\,\text{m} \end{bmatrix}$$

Study Questions

1. Given the geocentric coordinates for a place in Enterprise, AL on the NAD83 datum and the GPS baseline components from Enterprise to Banks, AL:

Station Enterprise	Vector
X = 392,159.8225 meters	ΔX = 1,539.3461 meters
Y = -5,437,468.8357 meters	ΔY = 26,752.1344 meters
Z = 3,299,781.6622 meters	ΔZ = 43,215.0193 meters

   a. What is the chord (mark-to-mark slant range) distance between the two points in miles?
   b. What are the geocentric coordinates of Banks using the GRS 80 ellipsoid?
   c. What are the geodetic coordinates of station Banks using the GRS 80 ellipsoid?

2. Compute geocentric coordinates for Point Zebra in meters on the GRS 80 ellipsoid. The known geographic (geodetic) coordinates for Point Zebra are:

   ϕ = 31° 27' 37.34298" N
   λ = 85° 59' 05.42009" W
   h = 123.651 meters.

3. Compute the geodetic coordinates for point Able on the GRS 80 ellipsoid. The known geocentric coordinates of Point Able are:

   X = -115,118.547 meters
   Y = -5,201,931.296 meters
   Z = 3,677,897.961 meters.

4. Given the geocentric coordinates for a place in Enterprise, AL in study problem #1. Compute the GRS 80 geodetic coordinates (ϕ, λ, h) for this place.

5. Point P is the origin of a local geodetic coordinate system. The geodetic coordinates (GRS 80) of point P are: λ = 120° W, φ = 60° N, h = 500 m. Point R having geodetic coordinates: λ = 120°10'00"W, φ = 59°12'30"N, h = 550 m is used as a backsight. The following field data was observed from point P to a foresight at point Q (assume that the height of the instrument at P equals

124

the height of the prism at Q, neglect earth curvature effects): Horizontal angle right = 248°08'12", Slope distance = 438.094 m, Zenith angle = 84°28'17".

    a. What are the geocentric coordinates of point P?

    b. Compute the geodetic azimuth and geodetic distance from P to R (you may use the NGS program *INVERSE* found at www.ngs.noaa.gov).

    c. What are the local geodetic coordinates of point Q?

6. The following table of points are defined in a 2-D coordinate system (x, y). If the coordinate axes are translated -5.25 in the x-direction and 21.05 in the y-direction, and rotated 135°, what are the coordinates of the points in the new system (x' y')? (Assume a scale factor of 1.00).

Point	x	y
A	0.00	0.00
B	-5.25	21.05
C	25.00	25.00
D	-10.00	-30.00
E	8.00	-12.00

7. The following table of points is defined in a 3-D coordinate system (x, y, z) which is graduated in feet. The origin of the **X** system has coordinates (-3.28, 62.35, 27.02) in another 3-D coordinate system (x', y', z') which is graduated in meters. The **X₁** system has been rotated relative to the **X** system by the following rotations in the order given: 15° about the z-axis, 235° about the x-axis, and -35° about the y-axis. Determine the coordinates of the points in the **X₁** system.

Point	x (ft)	y (ft)	z (ft)
A	0.000	0.000	0.000
B	100.000	100.000	100.000
C	-1237.540	642.020	-222.720

8. A total station is set up at station MEDIAN 2 which has GRS 80 geodetic coordinates: 121°49'25.98594" W, 42°15'15.61025"N, 1289.858 m. Three stations are observed with the following results:

Station	Geodetic Azimuth	Slant Range	Zenith Angle
A	63°25'42"	538.254 m	84°35'44"
B	313°52'33"	1042.836m	93°02'28"
C	203°12'02"	284.663 m	89°53'40"

Determine the geodetic coordinates of stations A, B and C to the same precision.

9. Compute the original system coordinates for the origin of the transformed system in Problem 7.9.

10. Determine the (x', y') coordinates of point C.

Point	x (ft)	y (ft)	x' (m)	y' (m)
A	0.00	0.00	16.179	6.382
B	10.37	56.11	0.000	0.000
C	29.17	-24.46	?	?

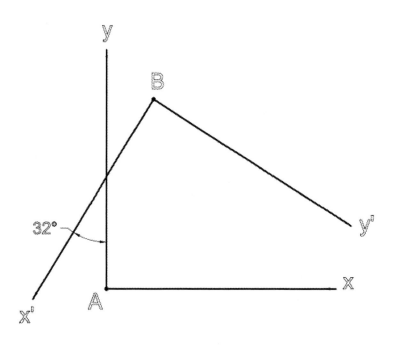

11. Given the following GRS 80 geodetic coordinates for points E and F, compute the geodetic azimuth from E to F to the nearest second.

Point	λ	φ	h (m)
E	120°00'00.00000"W	60°00'00.00000"N	1000.000
F	120°15'00.00000"W	60°06'00.00000N	1500.000

12. A point has coordinates (u, v, w) in an orthogonal, right-handed, three-dimensional coordinate system. The same point has coordinates (a, b, c) in another orthogonal, right-handed, three-dimensional coordinate system. The origin of the u-v-w system is known in the a-b-c system. The a-b-c system may be aligned with the u-v-w system using the following rotations in the order given: θ about the c axis and ω about the b axis. Write the symbolic expression for the u coordinate (i.e., u = ?) as a function of the rotation angles, translations and scale factor.

126

CHAPTER EIGHT

GEODETIC DATUMS

Clarification:

A horizontal datum is much more than just an ellipsoid. Practically speaking, a datum is a set of physical monuments that you can find on the ground, a reference surface, and a set of published coordinates. Further, a datum is defined by how the reference surface (ellipsoid) is fitted to the physical earth. Additionally, the coordinates on each monument are the result of a process consisting of observations (measurements), the reduction of these observations to geodetic quantities, and least squares adjustment(s) to provide the best fit of all measured quantities into a geodetic control network.

A *datum* is *any quantity or set of such quantities that may serve as a referent or basis for calculation of other quantities.* (NGS, 1986). A *geodetic datum* is *a set of constants specifying the coordinate system used for geodetic control.* (NGS, 1986) Historically, datums were defined by individual countries or perhaps a group of countries within a region. The modern trend is toward global datums that may be used worldwide. This chapter will focus on datums in use within the United States.

The Office of National Geodetic Survey located within the U.S. Department of Commerce is responsible for providing geodetic reference systems for use in the United States. This agency has enjoyed a long history resulting in several name changes in the past:

1807 Survey of the Coast
1836 U.S Coast Survey
1871 U.S. Coast and Geodetic Survey with a Geodesy Division
1970 National Ocean Survey with a Geodesy Division
1982 National Ocean Service with a Geodesy Division called the National Geodetic Survey Division.

These various names in history for the Office of National Geodetic Survey (NGS) are found on the face of geodetic control monuments found in survey practice.

Vertical and horizontal geodetic control datums have evolved separately through history because each use different reference surfaces. Mean sea level or the geoid have been the reference surfaces for the provision of elevations and orthometric heights in the vertical datum. The ellipsoid of revolution has been the reference surface for the provision of horizontal positions. This has encouraged the classification of vertical datums as one-dimensional (elevation only) and horizontal datums as two-dimensional (latitude and longitude). The development of a separate vertical datum has resulted in separate vertical control monuments for elevations called "benchmarks." Many of these benchmarks were established to accommodate the placement of a level rod, but not the setting of a tripod.

The civilian use of the Global Positioning Service (GPS) has brought a fundamental shift in the provision and use of geodetic control datums. It is now advantageous to occupy control stations with precise horizontal as well as vertical positions. As current and future geodetic research continues to bring advances in the provision of precise GPS-derived orthometric heights, it is not difficult to predict the eventual combination of vertical and horizontal datums.

The impact of GPS on geodetic datums is explored further in Chapter 12 GPS and Geodesy. In this chapter, we consider those horizontal and vertical control datums in current use and of the recent past in the United States.

Vertical Control Datums

A vertical control datum is a set of fundamental elevations to which other elevations are referred. (NGS, 1986). Orthometric heights are referred to a vertical control datum. Traditionally, orthometric heights (elevations) have been referred to mean sea level. This reference surface has caused numerous problems throughout history due to its ambiguous nature, that is, the equipotential surfaces of local mean sea level can vary as much as two meters from place to place. The reference surface of choice for the vertical datum is the geoid.

Vertical Control Datums in the U.S.

The first leveling route of geodetic quality in the United States was surveyed during 1856-57 by the U.S. Coast Survey in conjunction with a study of ocean currents and tides in New York Bay and the Hudson River. However, the "official" beginning of geodetic leveling was in 1887 at benchmark "A" in Hagerstown, Maryland. The leveling followed the triangulation arc previously established along the 39th parallel. The datum was mean sea level as determined by five tide stations along the Atlantic Ocean. By the year 1900 approximately 21,000 km of geodetic leveling had been completed in the U.S.

The addition of tide stations and expansion of the leveling network necessitated the periodic readjustment of the network. The U.S. Coast and Geodetic Survey performed readjustments in 1903 (8 tide stations, 32,000 km), 1907 (8 tide stations, 38,000 km) and in 1912 (9 tide stations, 46,000 km). By 1929 the first order leveling networks in the U.S. and Canada had been connected at 24 locations from the Atlantic coast to the Pacific coast. The U.S. network consisted of 21 tide stations and 75,000 km of network. The Canadian network consisted of 5 tide stations and 32,000 km of network. Another readjustment was performed in 1929.

National Geodetic Vertical Datum of 1929

The 1929 readjustment defined the *Sea Level Datum of 1929*. This datum was not adopted by Canada. The name of the datum was changed in 1973 to *National Geodetic Vertical Datum of 1929* (NGVD 29). A major problem in relying on "mean sea level" for the datum was that mean sea level varies by approximately 0.7 m from coast to coast and approximately 0.1 m along each coast (local mean sea levels are not equal!). In other words, the "mean sea level" tide stations were located on different equipotential surfaces. In addition to the distortions caused by local mean sea level variations, the NGVD 29 datum suffered distortions from crustal upheaval and subsidence, systematic errors, and deficiencies in network design.

North American Vertical Datum of 1988

In 1978 work was begun by the National Geodetic Survey to undertake a least squares adjustment of the entire North American geodetic leveling network. The result would be the *North American Vertical Datum of 1988* (NAVD 88). The adjustment included over 600,000 benchmarks across the continent and the re-leveling of 83,000 km of first order network. NAVD 88 facilitated the development of an improved geoid model that is instrumental in performing *GPS leveling*. Elevation differences between NGVD 29 and NAVD 88 are a maximum of approximately ±1 m in an absolute sense and approximately a few mm in a relative sense. Figure 8.1 shows the absolute datum shift from NGVD 29 to NAVD 88 (in mm).

It is important to realize that NAVD 88 is not referenced to mean sea level. In fact, the datum was obtained by fixing a single datum point, Father

Point/Rimouski, on the Great Lakes in Quebec, Canada. The National Geodetic Survey estimates that NAVD 88 suffers an offset of 31.4 centimeters from the Earth's best-fit global geopotential. This bias is considered in the development of the software product GEOID99 that produces interpolated geoid heights.

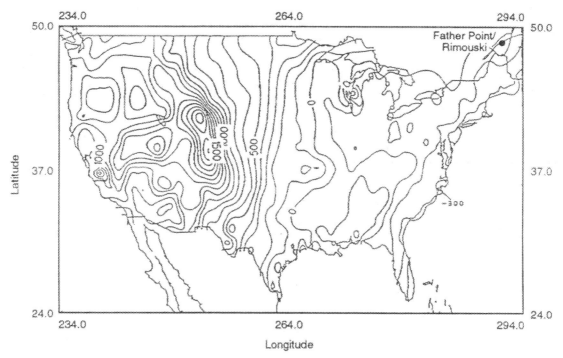

Figure 8.1 Datum shift between NGVD29 and NAVD88 in millimeters. (Zilkoski, Richards, and Young, 1992).

Horizontal Control Datums

Two major horizontal geodetic datums are used in the United States. The North American Datum of 1927 (NAD 27) was the current datum until 1986 when published coordinates for the North American Datum of 1983 (NAD 83) became available.

Published coordinates under the two datums for the same point will differ significantly by an amount termed the "datum shift." The datum shift from NAD 83 to NAD 27 is best illustrated in figures 8.2 and 8.3. There is no direct mathematical transformation between the two datums. Software programs can approximate the datum shift for mapping-grade work. The practicing professional is advised to include control monuments common to both datums in geodetic projects if output is desired in both datums. Three critical differences exist between the two datums:

130

1. The two datums use two different ellipsoids with different defining parameters. In fact, the two ellipsoids not only have different parameters, but also were constructed using different objectives and measurements. The Clarke Spheroid of 1866 (Clarke 1866), the ellipsoid used in NAD 27, was constructed from estimates of the earth's shape and size computed from data collected by arc measurements of the earth's surface. The Geodetic Reference System of 1980 (GRS 80), the ellipsoid used in NAD 83, was developed using the principles of physical geodesy outlined in Chapter Four to develop a "level ellipsoid" that closely approximates the geoid.

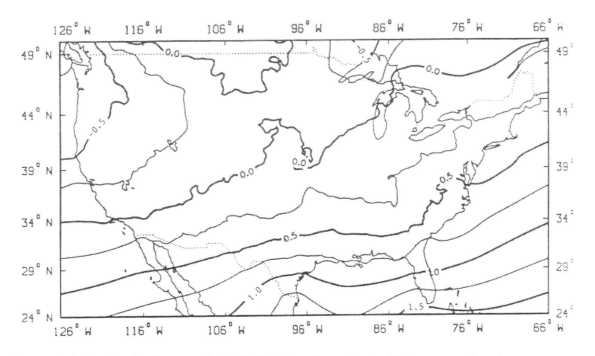

Figure 8.2 Latitude datum shift (NAD 83 minus NAD 27) in seconds of arc (NOAA Professional Paper NOS 2, Page 241).

2. The two ellipsoids were fitted to each datum in a different way. The Clarke 1866 ellipsoid was fitted to the regional or local geoid in the vicinity of the United States without regard to how the ellipsoid fit the geoid in the rest of the world. The GRS 80 ellipsoid was fitted to the center of the earth's mass providing a global datum.

3. NAD 27 was created by a partitioned least squares adjustment. Subsequent densification of the large triangulation loops provided known and suspected distortions of the geodetic control network. NAD 83 was created with a huge

simultaneous least squares adjustment that included the significant
contribution of new geodetic measurements.

Figures 8.2 and 8.3 indicate that the datum shift between NAD 27 and NAD 83
varies with position. For instance in Alabama, the datum shift between NAD 27
and NAD 83 can be as great as 20 meters in latitude and 5 meters in longitude.
The datum shift in Oregon is approximately 20 meters in latitude and 130
meters in longitude.

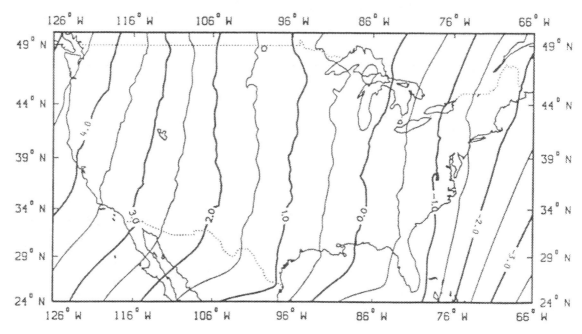

Figure 8.3 Longitude Datum Shift (NAD 83 minus NAD 27) in seconds of arc
(NOAA Professional Paper NOS 2, Page 242).

History

The need for a horizontal control network within the U.S. was addressed
when Congress and President Jefferson established the Survey of the Coast
through the Act of 1807. The purpose of this act was to map the shoreline,
islands, ports and other features of the Atlantic Seaboard. Two baselines were
measured in the New York - New Jersey area in 1816 for the start of
triangulation networks. In 1817 the first triangulation station WEASEL was
established in New Jersey. By 1843 the Eastern Oblique Arc consisted of 1200
stations.

Between 1871 and 1897 the 39th Parallel Arc (first order) was surveyed from the Atlantic coast to the Pacific coast along the route of the first transcontinental railroad. The second and third order arc extending along the Pacific coast from Canada to Mexico was completed in 1890. The Eastern Oblique Arc was completed by 1900. Figure 8.4 shows the extent of the first order horizontal control network within the U.S. in 1900.

North American Datum

The *United States Standard Datum* was formed from the New England Datum on March 13, 1901. The origin of this datum was Station PRINCIPIO in Maryland. The U.S. Standard Datum was extended and became known as the *North American Datum* in 1913 when it was adopted by Canada and Mexico for their control surveys. During World War I the primary method of extending control was the traverse due to the flatness of terrain surveyed, availability of highway and railroad corridors, and the prohibitive cost of constructing triangulation towers out of lumber. This latter problem was solved with the advent of the Bilby portable steel tower for triangulation in 1926.

Figure 8.4 Extent of Horizontal Geodetic Control in U.S. in 1900. (NOAA Professional Paper NOS 2, page 17).

North American Datum of 1927

By 1927 approximately 25,000 horizontal control stations had been established within the U.S. The network extent is shown in figure 8.5. Due to the way the network had expanded from a single triangulation arc into a network of triangulation and traverse arcs, significant distortions were present. A massive readjustment of the network was undertaken in 1927 to distribute errors and determine station positions. This adjustment resulted in the *North American Datum of 1927* (NAD 27).

Ewing, 1970 states that a regional datum may be defined with seven parameters: (1) a, the semimajor axis of the reference ellipsoid, (2) b, the semiminor axis of the reference ellipsoid, (3) ξ_0, the deflection of the vertical component in the meridian plane at datum initial point, (4) η_0, the deflection of the vertical component in the prime vertical at the initial point, (5) α_0, the geodetic azimuth from the initial point along an initial line of triangulation, (6) N_0, the geoid height at datum origin, and (7), the condition that the ellipsoid minor axis and the earth's rotation axis be parallel. (Note: deflection of the vertical will be discussed in Chapter 10.)

Figure 8.5 Extent of Horizontal Geodetic Control in U.S. in 1927. (NOAA Professional Paper NOS 2, page 18).

134

Parameters (3), (4), and (6) fix the relationship between geoid and ellipsoid at their common point at the datum initial point. If the geodetic coordinates at the initial point are assumed equal to the observed astronomic coordinates then:

$$\xi_0 = \Phi - \phi = 0 \quad and \quad \eta_0 = \Lambda - \lambda = 0$$

where Φ is astronomic latitude and Λ is astronomic longitude. Further, if the ellipsoid is fitted tangent to the geoid at the initial point then $N_0 = 0$. This would give a datum orientation that is based on a single astronomic position. While the initial point may have zero as the geoid height and zero for the components for the deflection of the vertical, the other triangulation stations in the datum will begin to show significant deflections. The fit of a regional datum causes the ellipsoid to meet the geoid at the initial point.

One last concern is the fact that the ellipsoid can rotate about its normal to the initial point tangent to the geoid. The ellipsoid can be rotated to bring its minor axis parallel to the rotation axis of the earth. Note, that in a regional datum, there is no requirement that the rotation axis of the earth and ellipsoid be coincident, only parallel.

Triangulation station MEADES RANCH in Kansas was chosen to be the initial point $\left(\Phi_0, \Lambda_0\right)$. The astronomic latitude and longitude of MEADES RANCH are $\Phi_0 = 39°\ 13'\ 25.67"$ N and $\Lambda_0 = 98°\ 32'\ 28.20"$ W. Station WALDO was chosen as the azimuth mark from the initial point in order to fix the orientation of the network. The initial azimuth from MEADES RANCH to WALDO was published as $\alpha_0 = 75°\ 28'\ 09.64"$.

The reference ellipsoid was chosen to be the *Clarke spheroid of 1866* (spheroid is synonymous with ellipsoid, not sphere!). This ellipsoid, having defining parameters $a = 6,378,206.4$ meters and $b = 6,356,583.8$ meters was chosen because it had been in usage for triangulation for the U.S. Coast and Geodetic Survey, and it demonstrated a close fit with the regional geoid.

Swartz 1989 makes several important points about the design of NAD 27. Numerous existing triangulation stations were examined to determine which set of coordinates would result in the sum of the squares of the differences between the astronomic and geodetic latitudes and longitudes being a minimum (this shows a design direction away from a single astronomical position type of orientation to that of an astrogeodetic orientation). Dr. John F. Hayford, in charge of the geodetic work in the United States during the development of NAD 27, selected MEADES RANCH as the initial point on this basis. The orientation of the Clark 1866 ellipsoid was controlled through the use of Laplace azimuths distributed throughout the triangulation network.

A Laplace Station is a triangulation station where astronomic latitude, longitude, and azimuth are known. As the triangulation is run on the earth's surface the computed geodetic coordinates at a Laplace Station can be compared to the observed astronomic coordinates to give an indication of closure:

$$\Lambda - \lambda = \eta \sec \phi$$

$$A - \alpha = \eta \tan \phi$$

$$\text{so} \quad A - \alpha = (\Lambda - \lambda) \sin \phi$$

$$\text{or} \quad (A - \alpha) - (\Lambda - \lambda) \sin \phi = 0$$

Therefore, the Laplace equation can be used to measure the accuracy of the differences: $(A - \alpha)$ and $(\Lambda - \lambda)$. The closure error (w):

$$(A - \alpha) - (\Lambda - \lambda) \sin \phi = w$$

is caused by errors in the astronomically observed and geodetically computed longitude and latitude. This type of datum orientation is known as astrogeodetic where the sum of the squares at of the components of the deflection of the vertical at Laplace Stations is held to a minimum.

The net result may be termed a "modified astrogeodetic" orientation of the ellipsoid for the NAD 27. Dr. Hayford's selection of the initial point to minimize the deflections at the Laplace Stations shows a desire for an astrogeodetic orientation, but the deflections and the geoid height at the initial point were intended to be zero. Heiskanen and Meinesz, 1956 state that independent absolute gravimetric measurements found the actual components of the defection of the vertical at Station Meades Ranch to be $\xi_0 = -1.3"$ and $\eta_0 = 0.3"$; not their intended values of zero.

Heiskanen and Meinesz emphasize the difference between gravimetrically measured deflections (See Chapter 9) and astrogeodetic deflections. The tool of geometric geodesy is triangulation with astronomic observations (the arc measurement method). The values of N (geoid height) and the components of the deflection of the vertical at triangulation stations depend on the dimensions of the reference ellipsoid used and on the components of the deflection of the vertical ξ_0 and η_0 assumed to exist at the initial point. Therefore the computed astrogeodetic deflections in a regional datum at other triangulation stations are relative and not necessarily in agreement with those found gravimetrically using the gravity field of the earth. It is important to see that astrogeodetic deflections and gravimetrically determined deflections are one of the important differences between a regional datum such as NAD 27 and a global datum such as NAD 83.

The final step in producing a regional datum is the least squares adjustment to obtain the 'best-fit' of redundant observations to produce published coordinates on the monuments. The actual adjustment method chosen was analogous to adjusting a level network due to the impracticality of performing a rigorous least squares adjustment. A least squares adjustment would have required the solution of 3,000 simultaneous linear equations. While this was possible it was not considered economical. The adjustment, which broke the network into halves and further subdivided it into loops (see figure 8.6), required five years to complete!

Figure 8.6 Least squares Adjustment for NAD 27. (NOAA Professional Paper NOS 2, Page 6).

The large triangulation loops that formed the datum in 1927 had experienced a densification of control within the loops during later years. This provided the problem of forcing new control work into existing triangulation loops. Improvements in surveying methods and instruments led to a need for a network where the relative precision between control monuments was one part in 100,000. NAD 27 was designed to provide a relative precision of one part in 25,000. The distortion provided by the original lack of a simultaneous least squares adjustment and the subsequent densification had degraded network precision to less than one part in 15,000 in places. The surveying and mapping community could easily detect this distortion with use of the new technology of the 1960's and 1970's. The huge technological advance was the new Electronic Distance Measuring Instrument (EDMI) that could measure precise distances with electromagnetic energy.

In addition to demands of the surveying and mapping community, other users began to ask for global geodetic reference data. The needs of growing cities, the military, and the space program had made NAD 27 obsolete.

The principal problems of NAD 27 were: (1) the lack of a simultaneous least squares adjustment, (2) the lack of a geoid model (since the ellipsoid was purposefully fitted to the regional geoid); (3) an insufficient number of baselines in the triangulation due to the prohibitive cost of precise taping, (4) the densification of the large triangulation loops, and (5) the need for a global instead of a regional datum.

In hindsight, it is easy to see that the NAD 27 served surveyors well from 1927 into the mid-1970's. At that time, the prohibitively high cost of determining precise baselines for the ordinary control survey was solved with the widespread use of the Electronic Distance Measuring Instrument (EDMI). The EDMI made it possible to develop a minimally constrained local project network that possessed greater network precision than was possessed between nearby NAD 27 control points. When surveyors could see the degradation of project precision gained by constraining the project to NAD 27 control points, the need for a new datum was persuasive.

North American Datum of 1983

By 1969 users within the U.S. and Canada were demanding a new adjustment of the horizontal control network within North America. The National Geodetic Survey initiated a new datum project in 1974 that resulted in the *North American Datum of 1983* (NAD 83). This datum is used in the Americas, from Canada to Central America.

NAD 83 is a global datum. One benefit of a global datum is that a point on one continent can be precisely tied to points on other continents . The development of a global geodetic datum requires the use of an ellipsoid that best fits (in a least squares sense) the entire geoid, not just a regional portion of the geoid such as that portion under the United States. The ellipsoid chosen for a global datum has a different design than ellipsoids used for regional datums. Remember the ellipsoid chosen for a regional datum tends to give the best fit to a local portion of the geoid that can be determined by geometric geodesy. An ellipsoid chosen for a global datum tends to provide the best fit to the entire geoid.

The concepts of physical geodesy are used to define an ellipsoid for use in a global datum. The earth was described in terms of gravity and potential in Chapter Four. An ellipsoid can be described in the same way. Let the mass of the ellipsoid rotate with the earth forming a gravity field termed the "Normal Gravity Field." Therefore, the surface of the ellipsoid is defined to be a level surface formed by its own gravity field. This level surface or "level ellipsoid" is defined by four parameters: $U = f(a, J_2, GM, \omega)$ where a is the semimajor axis, J_2 is the dynamical form factor, GM is the product of the gravitational constant times the mass of the ellipsoid, and ω is the angular velocity of the earth.

The International Union of Geodesy and Geophysics (IUGG) first defined an *International Ellipsoid* in 1924. This same body proposed the *Geodetic Reference System of 1967* (GRS 67) and most recently the *Geodetic Reference System of 1980* (GRS 80). The NAD 83 uses the GRS 80 ellipsoid. The defining parameters of GRS 80 are:

$a = 6{,}378{,}137$ meters

$$GM = 3.986005 \times 10^{14} \, m^3\!\!\Big/\!\text{sec}^2$$

$J_2 = 0.00108263$

$\omega = 7.292115 \times 10^{-5} \ radians/\sec$

The geometric flattening, f, of the ellipsoid is computed using these defining parameters.

The level ellipsoid is fitted to the earth. The initial point will not be a point on the geoid, but a point at the center of the earth's mass. Moritz, 1978 gives the following defining parameters for a global geodetic datum: (1) datum origin at the center of the earth's mass, (2) the z axis is in the direction of the Conventional International Origin (CIO) defining a mean North Pole (see Chapter 11), (3) the x axis is parallel with the Greenwich mean astronomical meridian which is by convention the starting point for the measurement of longitude, and (4) an ellipsoid defined by a, J_2, GM and ω as defined in the previous paragraph.

Note that a three dimensional Cartesian coordinate system (x, y, and z) centered at the datum origin can be logically defined in a global datum. In this sense, a global datum is three-dimensional. However, it must be remembered that the z coordinate is not yet directly convertible into a precise orthometric height desired by engineers to predict the flow of water. As discussed in Chapter Seven:

$$(x,y,z) = f(\lambda, \phi, h_e)_{a,f}$$

Note that the transformation provides ellipsoid height. The distance between the geoid and the ellipsoid surface (N) must be known to compute the orthometric height (H) where $H = h - N$. This topic is discussed further in Chapter Nine.

The least squares adjustment that led to NAD 83 was completed in 1986. The simultaneous least squares adjustment consisted of 266,000 stations and 1,785,000 observations (terrestrial, TRANSIT, and VLBI). The adjustment required 940 hours of computer time!

Two major global geodetic datums are in use today, NAD 83 and the *World Geodetic System, 1984*. The Defense Mapping Agency (DMA) now known as the National Imagery and Mapping Agency (NIMA) of the U.S. Department of Defense developed the World Geodetic System to track satellites and missiles and to provide a common reference for mapping, charting and geodetic products. The first datum was the *World Geodetic System of 1960* (WGS 60). Subsequent datum revisions were implemented in 1966 and 1972. The current datum is the *World Geodetic System of 1984* (WGS 84) which is the GPS satellite datum. Note that in the World Geodetic System the datum and the "earth-centered" or level ellipsoid have the same name. Fortunately, the two ellipsoids: GRS 80 and WGS 84 are nearly identical. The difference lies in the number of significant digits chosen for J_2, the dynamical form factor and, thus, slightly different geometric flattenings. Since the two ellipsoids were fitted to their respective datums identically, the two datums are also nearly identical. The positional variation of coordinates computed using these different datums is expected to be in the sub-millimeter range.

Table 8.1 Geometric ellipsoid parameters.

Datum	a	f
NAD 83	6,378,137.0 m	1/298.257222101
WGS 84	6,378,137.0 m	1/298.257223563

Geodetic Control Networks

Geodetic control networks provide the physical representation of a datum. Stations have historically been monumented using concrete monuments with brass tablets or brass tablets set in bedrock. The former type of monument is relatively unstable while the stability of the bedrock monument is typically unrivaled. Where bedrock is unavailable in modern geodetic surveys, monuments are constructed of a steel alloy rod driven to refusal and set in a sleeve that minimizes the effect of frost heave on the monument's position.

The NGS maintains a database of geodetic control stations (vertical and horizontal) and observations (terrestrial, astronomic, gravity, satellite, VLBI) within the U.S. This database is know as the *National Spatial Reference System* (NSRS). The control station portion of the database may be accessed via the Internet (www.ngs.noaa.gov/FORMS/ds_desig.html) or users may purchase a CD-ROM consisting of geodetic control stations within regions of the country. A *data sheet* has been compiled for each station within the database. The data sheet contains information about the station's position, monument type, location directions and other relevant information. A sample data sheet is included in Appendix A.

A significant problem with the NSRS lies with monumentation. Lack of adequate funding has led to dissolution of a maintenance program for NSRS monuments, many of which were established in the 1940's. Some of the monuments have been disturbed or destroyed due to construction. Because line of sight between stations was the paramount consideration in station selection, many stations are located at the top of hills and mountains. Also, stations are often located on private land. These qualities often translate into an accessibility problem for everyday use by surveyors.

Perhaps the most conspicuous inadequacy of the existing NSRS stations is their lack of suitability for GPS observations. Line of sight to or from a triangulation station was only required along a few discrete corridors. Obstructions around a station which did not inhibit the view of other stations were not a concern. The ideal station for GPS observations is free from obstructions from 0° to 360° in azimuth and from 15° above the horizon to the zenith. The ability to see other network stations is not a consideration. Many NSRS stations fail to meet these requirements for GPS observations.

High Accuracy Reference Networks

The usage of GPS in geodetic control surveying highlighted the deficiencies within the pre-GPS NSRS. The 270,000 horizontal control stations that were observed by high accuracy classical means (triangulation, traversing, trilateration) possess relatively low accuracy positions with respect to GPS.

Table 8.2 Classical Horizontal Control Network Accuracy Standards (FGCC 1984)

Classification	Ratio	PPM
First Order	1:100,000	10
Second Order		
Class I	1:50,000	20
Class II	1:20,000	50
Third Order		
Class I	1:10,000	100
Class II	1:5,000	200

Table 8.2 contains the classical accuracy standards for horizontal control networks developed by the Federal Geodetic Control Committee (FGCC). The FGCC has also developed preliminary standards for horizontal control surveys performed using GPS or other space system techniques. Table 8.3 provides a listing of these accuracy standards. The base error component shown in Table 8.3 is that error present in the measuring scheme that is independent of the line length. Tribrach centering errors would be included in this component.

The accuracy standards reveal that the potential accuracies possible with GPS surveying techniques are several magnitudes of order higher than those possible with classical surveying techniques. This has little effect on the property surveyor, but it is of significant interest to geodetic control surveyors. Extremely high accuracy control surveys may now be performed with a fraction of the effort that was required to perform triangulation.

Table 8.3 GPS Horizontal Control Network Accuracy Standards (FGCC 1989)

Order	Base Error	Line-Length Dependent Error
AA	0.3 cm	1:100,000,000 (0.01 ppm)
A	0.5 cm	1: 10,000,000 (0.1 ppm)
B	0.8 cm	1: 1,000,000 (1 ppm)

Suppose two-second order (1:50,000) NSRS stations are used to control a GPS network. In all likelihood, the baseline vector connecting the two known stations will have a slightly different length than that determined by inversing using the published station coordinates. The accuracy of the GPS vector will probably be on the order of 1:500,000 if single frequency survey grade receivers are used. By constraining the position of the two-second order stations the higher accuracy GPS derived baseline vectors will be distorted to fit into the existing control framework. This distortion is undesirable, effectively corrupting accurate baselines. Ideally, network densification is accomplished within a higher accuracy framework.

A *High Accuracy Reference Network* (HARN) is a geodetic network of stations established using GPS observations. Early HARNs were known as *High Precision Geodetic Networks* (HPGN). A HARN is usually a state-based network

143

consisting of regularly spaced stations. The spacing between HARN stations is typically 25 to 100 km (16 to 62 miles). Ties are made to selected existing NSRS stations, nearby existing HARN stations (if available), and the global *Cooperative International GPS Network* (CIGNET).

Figure 8.7 shows the original Oregon HPGN (HARN), observed in 1990. The Oregon HARN was designed such that the station spacing would be approximately 25 km west of the Cascade Mountains and 50 km east of the Cascades. Note that Station ALTAMONT (ALTA) is located near the southern border of Oregon. The partial data sheet for Station ALTAMONT is contained in figure 8.8.

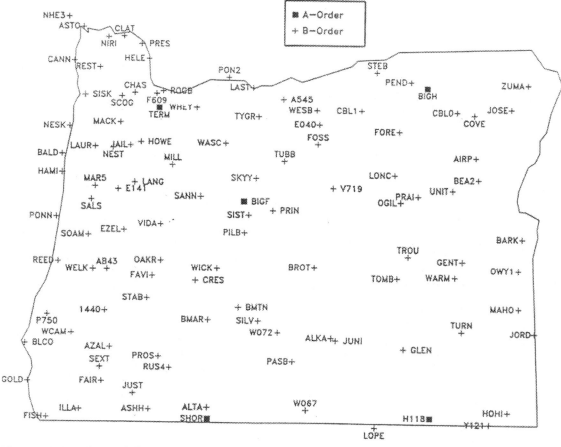

Figure 8.7 The original Oregon High Precision Network.

The NGS Data Sheet
****
NY0977 FBN   - This is a Candidate for Federal Base Network Control.
NY0977 DESIGNATION - ALTAMONT
NY0977 PID   - NY0977
NY0977 STATE/COUNTY- OR/KLAMATH
NY0977 USGS QUAD   - ALTAMONT (1995)
NY0977
NY0977   *CURRENT SURVEY CONTROL
NY0977 _____
NY0977* NAD 83(1998)- 42 12 32.56871(N) 121 44 50.17194(W) ADJUSTED
NY0977* NAVD 88   -   1250.2 (meters) 4102. (feet) GPS OBS
NY0977
NY0977 _____
NY0977 X   - -2,490,031.243 (meters)   COMP
NY0977 Y   - -4,024,274.156 (meters)   COMP
NY0977 Z   - 4,263,655.957 (meters)   COMP
NY0977 LAPLACE CORR-   2.97 (seconds)   DEFLEC99
NY0977 ELLIP HEIGHT-   1227.56 (meters)   GPS OBS
NY0977 GEOID HEIGHT-   -22.62 (meters)   GEOID99
NY0977
NY0977 HORZ ORDER - B
NY0977 ELLP ORDER - THIRD   CLASS I
NY0977
NY0977.The horizontal coordinates were established by GPS observations
NY0977.and adjusted by the National Geodetic Survey in July 1997.
NY0977.This is a SPECIAL STATUS position. See SPECIAL STATUS under the
NY0977.DATUM ITEM on the data sheet items page.
NY0977
NY0977.The orthometric height was determined by GPS observations and a
NY0977.high-resolution geoid model.
NY0977
NY0977.The X, Y, and Z were computed from the position and the
NY0977.ellipsoidal ht.
NY0977
NY0977.The Laplace correction was computed from DEFLEC99 derived
NY0977.deflections.
NY0977
NY0977.The ellipsoidal height was determined by GPS observations
NY0977.and is referenced to NAD 83.
NY0977
NY0977.The geoid height was determined by GEOID99.
NY0977
NY0977;   North   East   Units Scale   Converg.
NY0977;SPC OR S - 61,017.921 1,397,001.991 MT 1.00003361 -0 51 11.9
NY0977;UTM 10 - 4,673,746.444 603,408.677 MT 0.99973157 +0 50 30.1
NY0977
NY0977   SUPERSEDED SURVEY CONTROL
NY0977
NY0977 NAD 83(1991)- 42 12 32.56785(N) 121 44 50.17053(W) AD(   ) B

145

NY0977 ELLIP HT - 1227.63 (m)   GP(   ) 4 1
NY0977 NGVD 29 - 1249.16 (m)   4098.3   (f) LEVELING   3
NY0977
NY0977.Superseded values are not recommended for survey control.
NY0977.NGS no longer adjusts projects to the NAD 27 or NGVD 29 datums.
NY0977.See file dsdata.txt to determine how the superseded data were
NY0977.derived.
NY0977
NY0977_MARKER: DD = SURVEY DISK
NY0977_SETTING: 7 = SET IN TOP OF CONCRETE MONUMENT
NY0977_STAMPING: ALTAMONT
NY0977_MARK LOGO: ORDT
NY0977_MAGNETIC: N = NO MAGNETIC MATERIAL
NY0977_STABILITY: C = MAY HOLD, BUT OF TYPE COMMONLY SUBJECT TO
NY0977+STABILITY: SURFACE MOTION
NY0977_SATELLITE: THE SITE LOCATION WAS REPORTED AS SUITABLE FOR
NY0977+SATELLITE: SATELLITE OBSERVATIONS - November 13, 1997
Figure 8.8 NGS Data Sheet for station ALTAMONT (partial listing).

A HARN is three-dimensional in the sense that precise ellipsoid heights are determined by redundant network observations and adjustment. Accurate orthometric heights for the HARN stations can only be determined by precise leveling from existing vertical control stations. Typically second or third order leveling procedures are used to determine elevations for some or all of the HARN stations. This additional information allows accurate determination of the geoid at these HARN stations.

The HARN were established minimally to B-order accuracy (8 mm + 1:1,000,000) using dual frequency geodetic grade GPS receivers in order to provide a suitable control network for future GPS (and other) surveys. The stations were chosen such that they are easily accessible and provide no obstructions to GPS observations.

GPS surveys within the HARN require minimal adjustment to observations in order to orient the survey within the HARN. The network provides an ideal framework for Land Information Systems (LIS) and Geographic Information Systems (GIS) since all data may be referenced to a common system. This minimizes "fit" problems between data obtained from different sources. The use of independent project datums is no longer necessary.

The recent completion of the Indiana HARN sets the stage for a national adjustment of the individual HARNs into a national network. Many would be tempted to lightly dismiss the role of NAD 83 in light of the more precise coordinates on state HARN control monuments. Remember the datum for each individual HARN is still NAD 83. In truth, it can be said that HARN points possess more precise NAD 83 coordinates.

The NGS has developed a method of denoting the adjustment date associated with NAD 83 coordinates. The original NAD 83 adjustment is denoted NAD 83 (1986). HARN adjustments of NAD 83 are also dated. Coordinates based upon the original Oregon HARN adjustment are denoted NAD 83 (1991).

Reobservation of HARNs

The first HARNs were observed primarily with the goal of producing highly accurate horizontal (latitude and longitude) positions with vertical positions (ellipsoid height) being of less importance. Consequently, relatively few benchmarks were included in the first HARNs. However, the desire to determine accurate orthometric heights using GPS grew tremendously during the development and observation of subsequent HARNs. The growing *height modernization* focus of the National Geodetic Survey led to the inclusion of greater percentages of benchmarks in the later-observed HARNs.

It became apparent to the NGS and HARN users that additional benchmarks were needed in the first HARNs in order to refine the geoid and allow more accurate orthometric height determinations using GPS. Consequently, the NGS, in conjunction with state agencies and private surveyors, performed reobservations of the early HARNs, beginning in 1998. These reobservations included most of the original HARN stations and additional benchmarks to form the *Federal Base Network* (FBN) within a state. Through the cooperative efforts of those within the state, additional stations were observed to form the *Cooperative Base Network* (CBN). Both FBN and CBN stations are observed during the same *campaign*. This partnership during the reobservation of the HARN allows users within the state to derive the maximum benefit from the reobserved HARN with a denser network of control positions. The reobserved Oregon HARN is shown in figure 8.9. Positions for the reobserved HARN stations are still expressed in NAD 83, although a date (e.g., NAD 83 (1998)) is used to specify the *realization* of the datum.

Geodetic Datum Conversions

It is often desirable to convert the coordinates of a point in one datum to another. This is possible in most cases, either by using software or by performing fieldwork. The most critical question in datum conversions is: How accurate is the conversion? While nearly any datum conversion is possible, how will the accuracy of the converted position(s) affect their use? Datum conversions should not be regarded as a substitute for field observations! Where datum conversions are incapable of producing desired accuracies, field observations should be made using suitable procedures.

Vertical Datum Conversions

The simplest form of a datum conversion, the conversion between two vertical datums consists of the determination of the vertical shift (translation) from one vertical datum to another. While this may seem like a simple task, and it can be, it can become rather complicated, depending upon the extent of the project area and the desired accuracy of the conversion.

NGS Methods

Surveyors working in the U.S. have most need of a vertical datum conversion that converts NGVD 29 elevations to NAVD 88 orthometric heights. The NGS suggests four methods for performing this conversion.

1. Estimation of benchmark elevations by incorporating the original leveling data into NAVD 88 using least squares adjustment techniques. This is the most rigorous technique. NGS will adjust and publish the results if the data are submitted to NGS in computer-readable form.

2. Rigorous transformation of benchmark elevations for a particular project using datum conversion correctors estimated from the project's original adjustment constraints and their differences between NAVD 88 and NGVD 29. This technique may meet many users' requirements, but depending upon the accuracy requirements and the complexity of the user's leveling network, may prove to be more time-consuming than technique #1.

148

3. A simplified transformation of benchmark elevations using an average bias shift for the area. This technique should be the easiest to implement, but in general is only sufficiently accurate enough to meet mapping requirements.

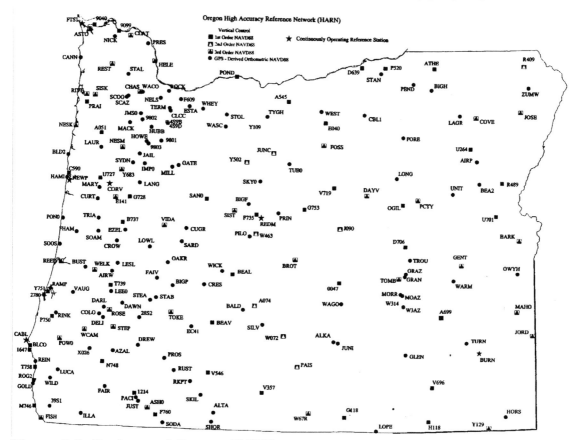

Figure 8.9 Reobserved Oregon HARN

Problem 8.1: You have been hired to determine the NAVD 88 orthometric height of benchmark T547 which has a NGVD 29 elevation of 4104.04 US ft. During your research you find benchmarks in the vicinity of OLD having the following published values.

Benchmark	NGVD 29 Elevation	NAVD 88 Orthometric Height
Q547	4088.82 US ft	1247.360 m
A15	4181.56 US ft	1275.636 m
AIRPORT 2	4085.32 US ft	1246.314 m
NORTH BASE	4191.80 US ft	1278.748 m

Convert the NGVD 29 elevations to meters and then compute the difference between the NAVD 88 orthometric heights. These differences are the "bias shift" values described in NGS method 3.

149

Benchmark	NGVD 29	NAVD 88	Bias Shift
Q547	1246.275 m	1247.360 m	1.085 m
A15	1274.542 m	1275.636 m	1.094 m
AIRPORT 2	1245.208 m	1246.314 m	1.106 m
NORTH BASE	1277.663 m	1278.748 m	1.085 m

The average bias shift equals 1.092 m. The estimated orthometric height for T547 is:

$$H_{T547} = 4104.04 \text{ US ft} \left(\frac{12.00 \text{ in / US ft}}{39.37 \text{ in / m}} \right) + 1.092 \text{ m} = 1250.914 \text{ m} + 1.092 \text{ m} = 1252.006 \text{ m}$$

The accuracy of this orthometric height is difficult to ascertain since no accuracy information regarding the four nearby benchmarks is provided. Suffice it to say that the prudent professional would not consider this computed value to be accurate to the millimeter. A conservative estimate is that the orthometric height of T547 is 1252.01 m ± 0.05 m.

4. Application of vertical shift value obtained from NGS *VERTCON* program. This method will give the user results having an *estimated* maximum error of ±2 cm (1σ).

Problem 8.2: Compute the NAVD 88 orthometric height for benchmark T547 using *VERTCON* from the NGS Web site. The output is below.

Latitude: 42 11 16.91546
Longitude: 121 44 12.10615
NGVD 29 Height: 4104.04 ft
Datum shift(NAVD 88 minus NGVD 29):   3.566 feet
Converted to NAVD 88 height: 4107.606 feet

The NAVD 88 orthometric height for T547 = 4107.61 US ft = 1252.00 m

The "Two-Plane" Method

Many cities, towns, counties, etc. have established local vertical datums to which elevations are referred. These add to the confusion of vertical datums! Conversions from these datums to a national datum (i.e., NGVD 29 or NAVD 88) cannot be accomplished by any of the NGS methods with the exception of #3. A generic method, termed the "two-plane" method, may be used to perform any vertical datum conversion.

The two-plane method may be used to estimate the vertical shift based upon horizontal position using a minimum of four points having known

150

elevations in two vertical datums. The method assumes each datum (level) surface is a plane surface, i.e., it neglects curvature of the earth. This assumption is valid for projects of limited extent. This method provides a least squares "best-fit" to determine the rotation angles between the two planes (rotation about N-axis and rotation about E-axis) and the vertical shift. Accuracy of shifted elevations is dependent on: the slope of the datum surfaces, the extent of the project area, and the terrain of the project area. The two-plane method is based upon the equation: Datum Shift$_i$ = α_E (N$_i$ - N$_0$) + α_N (E$_i$ - E$_0$) + t$_Z$ where: *Datum Shift$_i$* is the datum shift at point i; α_N and α_E represent rotations about the N and E axes, respectively; *N$_i$* and *E$_i$* are the plane coordinates (or estimates thereof) of point i; *N$_0$* and *E$_0$* are the plane coordinates of the centroid of the project area; and, *t$_Z$* is the vertical shift between the two planes at the centroid.

Once the user has solved for the three transformation parameters (α_N, α_E and *t$_Z$*) using least squares, the datum shift at any point within the project area may be estimated by inputting the coordinates of that point into the equation. The accuracy of this method may be best estimated by examining the residuals of the observations.

Problem 8.3: The data given in Problem 8.1, together with the horizontal positions, expressed using state plane coordinates, of the benchmarks is summarized below. Estimate the orthometric height of T547 using the two-plane method.

Benchmark	NGVD 29 Elevation	NAVD 88 Orthometric Height	N (m)	E (m)
Q547	4088.82 US ft	1247.360 m	60,320	1,395,020
A15	4181.56 US ft	1275.636 m	60,560	1,399,870
AIRPORT 2	4085.32 US ft	1246.314 m	56,300	1,397,560
NORTH BASE	4191.80 US ft	1278.748 m	57,867	1,401,028
T547	4104.04 US ft	?	58,670	1,397,840

The centroid of the project area is determined by computing the mean position of the four known benchmarks. Thus, N$_0$ = ΣN$_i$ and E$_0$ = ΣE$_i$. For this problem, N$_0$ = 58,762 m and E$_0$ = 1,397,840 m. As presented above, the functional model (equation) for this problem is:

Datum Shift$_i$ = (NAVD88$_i$ – NGVD29$_i$) = α_E (N$_i$ - N$_0$) + α_N (E$_i$ - E$_0$) + t_z

One equation may be written for each of the four known benchmarks (after simplifying):

Q547:	1.085 m = α_E (1558) + α_N (-3350) + t_z
A15:	1.094 m = α_E (1798) + α_N (1501) + t_z
AIRPORT 2:	1.106 m = α_E (-2462) + α_N (-810) + t_z
NORTH BASE:	1.085 m = α_E (-895) + α_N (2659) + t_z

The above equations are of the form $\mathbf{f} = \mathbf{B}\Delta + \mathbf{v}$ where $\mathbf{f}$ represents the vector of observations, $\mathbf{B}$ represents the matrix of parameter coefficients, Δ represents the vector of parameters and $\mathbf{v}$ represents the vector of observation residuals. The parameters for this system of equations may be estimated using unweighted least squares by the adjustment of indirect observations (Mikhail, 1976). The vector of parameters is computed using $\Delta = (\mathbf{B}^T\mathbf{B})^{-1}\mathbf{B}^T\mathbf{f}$.

$$\mathbf{B} = \begin{bmatrix} 1558 & -3350 & 1 \\ 1798 & 1501 & 1 \\ -2462 & -810 & 1 \\ -895 & 2659 & 1 \end{bmatrix} \qquad \mathbf{f} = \begin{bmatrix} 1.085 \\ 1.094 \\ 1.106 \\ 1.085 \end{bmatrix}$$

$$\Delta = \begin{bmatrix} \alpha_E \\ \alpha_N \\ t_z \end{bmatrix} = (\mathbf{B}^T\mathbf{B})^{-1}\mathbf{B}^T\mathbf{f} = \begin{bmatrix} -2.96861 \times 10^{-6} \text{ radians} \\ -5.72027 \times 10^{-7} \text{ radians} \\ 1.092 \text{ m} \end{bmatrix}$$

T547: Datum Shift = $(-2.96861 \times 10^{-6})(-92) + (-5.72027 \times 10^{-7})(-530) + 1.092 = 1.093$ m.
$\therefore H_{T547} = 4104.04$ US ft + 1.093 m = 1250.914 m + 1.093 m = 1252.006 m.

The accuracy of this value may be estimated by evaluation of the vector of observation residuals computed using $\mathbf{v} = \mathbf{f} - \mathbf{B}\Delta$. This vector describes the "fit" between the planes representing the NAVD 88 datum and the NGVD 29 datum.

$$\mathbf{v} = \begin{bmatrix} -0.005 \text{ m} \\ 0.008 \text{ m} \\ 0.006 \text{ m} \\ -0.009 \text{ m} \end{bmatrix}$$

Horizontal Datum Conversions

The frequent horizontal datum conversion problem faced in the U.S. is the conversion between NAD 27 and NAD 83. Unfortunately the distortions present in NAD 27 make this conversion complex and inaccurate. The NGS has developed the *NADCON* program to perform this conversion. *NADCON* uses a technique known as "minimum-curvature-derived surfaces" in order to perform the datum conversion. The accuracy of the *NADCON* datum conversion is approximately ±15 cm (1σ). Accordingly, the results of the conversion must be used with caution.

Problem 8.4: Station T547 has NAD 27 geodetic coordinates 42°11'17.38984"N, 121°44'08.10135"W. Compute NAD 83 (1986) coordinates for T547 using *NADCON*.

North American Datum Conversion
NAD 27 to NAD 83
NADCON Program Version 2.11

===

Transformation #:   1   Region: Conus

	Latitude	Longitude
	Latitude	Longitude
NAD 27 datum values:	42 11 17.38984	121 44 8.10135
NAD 83 datum values:	42 11 16.91546	121 44 12.10615
NAD 83 - NAD 27 shift values:	-0.47438	0.00480(secs.)
	-14.637	91.895 (meters)

Magnitude of total shift: 93.053 (meters)

Conversions between NAD 83 and NAD 83 HARN may also be performed using *NADCON*. The stated accuracy of this conversion is ±5 cm (1σ). The *NADCON* program is also contained in the *CORPSCON* program that has a "friendlier" user interface. The *CORPSCON* program may be downloaded from the NGS Web site.

Problem 8.5: Convert the NAD 83 (1986) position of station T547 into a NAD 83 (1991) position based upon the initial adjustment of the Oregon HARN.

North American Datum Conversion
NAD 83 to HPGN
NADCON Program Version 2.11

===

Transformation #:   1   Region: wohpgn

153

	Latitude	Longitude
NAD 83 datum values:	42 11 16.91546	121 44 12.10615
HPGN datum values:	42 11 16.91416	121 44 12.10060
HPGN - NAD 83 shift values:	-0.00130	-0.00555(secs.)
	-0.040	-0.127 (meters)

Magnitude of total shift: 0.134 (meters)

Another technique that may be used to perform the conversion between NAD 27 and NAD 83 is the four-parameter coordinate conversion discussed in Chapter Seven. Given a minimum of three points having known coordinates in both datums, a least squares adjustment may be used to estimate the four transformation parameters (one rotation, two translations, one scale change) needed to perform the conversion. Once the transformation parameters have been estimated the equations may be used to perform the datum transformation given point positions in the known datum. Accuracy of the conversion may be estimated by examining the residuals of the observations and is dependent on the project area and other factors.

Problem 8.6: It is desired to estimate the NAD 83 state plane coordinates for station T547 given the following information. Use a four-parameter transformation.

Station	NAD 27 State Plane Coordinates (US ft)		NAD 83 State Plane Coordinates (m)	
	Y	X	N	E
Q547	197,942.61	1,655,873.51	60,320.528	1,395,020.737
A15	198,728.51	1,671,786.31	60,560.076	1,399,870.961
AIRPORT 2	184,752.23	1,664,207.09	56,300.124	1,397,560.850
NORTH BASE	189,896.16	1,675,583.31	57,867.993	1,401,028.295
T547	192,530.32	1,665,125.05	?	?

Recall from Chapter Seven the form of the four-parameter transformation.

$$\begin{bmatrix} x' \\ y' \end{bmatrix} = \begin{bmatrix} a & b \\ -b & a \end{bmatrix} \begin{bmatrix} x \\ y \end{bmatrix} + \begin{bmatrix} c \\ d \end{bmatrix} = \begin{bmatrix} ax + by + c \\ ay - bx + d \end{bmatrix}$$

154

where a = s cos θ, b = s sin θ, c = $t_{x'}$ and d = $t_{y'}$. For this problem it is convenient to assume the (x, y) system as corresponding with the NAD 27 datum and the prime system with the NAD 83 datum. The two equations shown above may be written for each point known in both datums.

Q547:   1,395,020.737 = a(1,655,873.51) + b(197,942.61) + c
                60,320.528 = a(197,942.61) − b(1,655,873.51) + d
A15:   1,399,870.961 = a(1,671,786.31) + b(198,728.51) + c
                60,560.076 = a(198,728.51) − b(1,671,786.31) + d
AIRPORT 2:   1,397,560.850 = a(1,664,207.09) + b(184,752.23) + c
                56,300.124 = a(184,752.23) − b(1,664,207.09) + d
NORTH BASE: 1,401,028.295 = a(1,675,583.31) + b(189,896.16) + c
                57,867.993 = a(189,896.16) − b(1,675,583.31) + d

Since these equations are of the same form as the equations in Problem 8.3, we will solve for the four parameters using the unweighted adjustment of indirect observations [Mikhail, 1976]:

$$\mathbf{B} = \begin{bmatrix} 1,655,873.51 & 197,942.61 & 1 & 0 \\ 197,942.61 & -1,655,873.51 & 0 & 1 \\ 1,671,786.31 & 198,728.51 & 1 & 0 \\ 198,728.51 & -1,671,786.31 & 0 & 1 \\ 1,664,207.09 & 184,752.23 & 1 & 0 \\ 184,752.23 & -1,664,207.09 & 0 & 1 \\ 1,675,583.31 & 189,896.16 & 1 & 0 \\ 189,896.16 & -1,675,583.31 & 0 & 1 \end{bmatrix} \qquad \mathbf{f} = \begin{bmatrix} 1,395,020.737 \\ 60,320.528 \\ 1,399,870.961 \\ 60,560.076 \\ 1,397,560.850 \\ 56,300.124 \\ 1,401,028.295 \\ 57,867.993 \end{bmatrix}$$

$$\Delta = \begin{bmatrix} a \\ b \\ c \\ d \end{bmatrix} = \left(\mathbf{B}^T\mathbf{B}\right)^{-1}\mathbf{B}^T\mathbf{f} = \begin{bmatrix} 0.304799 \\ -1.34323 \times 10^{-6} \\ 890,311.767 \\ -14.489 \end{bmatrix}$$

N_{T547} = y'_{T547} = a(Y_{T547}) − b(X_{T547}) + d = 58,670.873 m

E_{T547} = x'_{T547} = a(X_{T547}) + b(Y_{T547}) + c = 1,397,840.621 m

Global Datum Conversions

Where two global datums share the same origin and orientation of axes but different ellipsoid parameters, such as GRS 80 (NAD 83) and WGS 84, the transformation of coordinates in one datum to the other may be accomplished using the form:

$$
\begin{bmatrix} \lambda \\ \phi \\ h \end{bmatrix}_{D_1} \leftrightarrow \begin{bmatrix} x \\ y \\ z \end{bmatrix} \leftrightarrow \begin{bmatrix} \lambda \\ \phi \\ h \end{bmatrix}_{D_2}
$$

Where the origin, orientation and ellipsoids differ, the general form must be used:

$$
\begin{bmatrix} \lambda \\ \phi \\ h \end{bmatrix}_{D_1} \leftrightarrow \begin{bmatrix} x \\ y \\ z \end{bmatrix}_{D_1} \leftarrow \text{Seven} - \text{Parameter Transformation} \rightarrow \begin{bmatrix} x \\ y \\ z \end{bmatrix}_{D_2} \leftrightarrow \begin{bmatrix} \lambda \\ \phi \\ h \end{bmatrix}_{D_2}
$$

Where the coordinate systems are *nearly* aligned, the seven-parameter transformation simplifies to:

$$
\begin{bmatrix} x \\ y \\ z \end{bmatrix}_{D_2} = (1+s) \begin{bmatrix} 0 & \gamma & -\beta \\ -\gamma & 0 & \alpha \\ \beta & -\alpha & 0 \end{bmatrix} \begin{bmatrix} x \\ y \\ z \end{bmatrix}_{D_1} + \begin{bmatrix} t_x \\ t_y \\ t_z \end{bmatrix}_{D_2}
$$

where s is the scale change, α, β, and γ are the differential rotation angles, expressed in radians, about the first, second and third axes, respectively. We will see application of the seven-parameter transformation for global datum conversions in Chapter 12.

Study Questions

1. Imagine yourself given the assignment to go out and establish a regional geodetic datum like NAD 27. List the seven defining parameters needed for such an effort.
2. What is the major problem faced in using a control point with NAD 27 coordinates in a project where the output is required in terms of the NAD 83 Geodetic Datum?
3. What is a level ellipsoid?
4. If the geoid height (N) at a place is negative, is the level ellipsoid above or below the geoid?
5. You find benchmarks that cannot be occupied by a tripod such as a brass cap set in a narrow bridge abutment or a railroad spike set in a brick wall. Explain why this was considered routine practice before GPS control surveys were possible?
6. Since GRS 80 is considered to be a level ellipsoid:
   a) Where is its center located with respect to the earth?
   b) Where is the z-axis of the xyz geocentric coordinate system located with respect to the level ellipsoid and the earth?
   c) Where is the x-axis of the xyz geocentric coordinate system located with respect to the level ellipsoid?

7. Discuss the factors that make NAD 27 coordinates so much different than NAD 83 coordinates for the same point on the earth's surface.
8. If a surveyor knows the ellipsoid height at a place and adds the geoid height, what does the sum represent?
9. Geographic coordinates were scaled off a USGS Topographic Quad map:

ϕ = 31° 58' 35"N; λ = 85° 02' 57"W

The horizontal datum upon which the map was compiled is given as NAD 27. You are working in NAD 83.
   a) Use the software program NADCON to convert the coordinates to NAD 83. Go to http://www.ngs.noaa.gov/TOOLS/Nadcon/Nadcon.html. Select Interactively compute a datum shift between NAD 27 and NAD 83.
   b) What is meant by datum shift?
   c) Verify the conversion of the datum shift values from arc seconds to meters.
   d) Use NADCON to convert the NAD 83 values to NAD 83 (HARN).

CHAPTER NINE

THE GEOID

As for the geoid, it is not a mathematical surface, but depends on the irregular distribution of visible and invisible masses of matter near the earth's surface. Hence, it must be determined by observations, point by point....to determine the absolute undulations of the geoid, we must use the gravimetric method. The mathematical basis of this method lies in the genius work of G.G. Stokes, published 102 years ago , in 1849. His formula for computing the undulations N has not required any essential correction.
     Heiskanen, 1951.

Now that we have explored geometric geodesy in excruciating detail, it is time to investigate what Torge calls *the geodetic boundary value problem.* We prefer to call it the connection between physical geodesy and geometric geodesy. This investigation will answer questions you may have like "How do the ellipsoid and geoid relate to one another?" We're glad you asked!

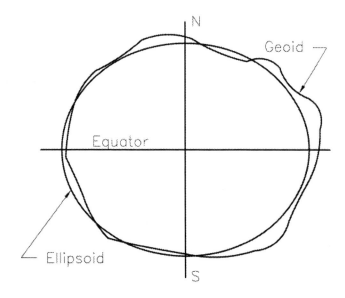

Figure 9.1 The geoid.

The student needs to realize that we are dealing with an earth-centered level ellipsoid such as GRS 80 fit to a global datum such as NAD 83. This chapter does not relate nor make sense applied to a geometrically defined ellipsoid such as the Clarke Spheroid of 1866 regionally fit to the local geoid for NAD 27 (See Chapter Eight). Now the finality of the big difference between NAD

27 and NAD 83 can be realized. NAD 27 has no geoid model. Since the age of satellite geodesy provides measurements to the ellipsoid, a way of converting ellipsoid heights to orthometric heights is critical. NAD 27 does not offer this conversion.

The Marine Geoid

The geoid, the fundamental reference surface of physical geodesy, was previously defined as an irregular surface that cannot be described with simple mathematical equations. Such a surface, like the earth's topography, must be discretely sampled point by point. Greater accuracy is generally gained by sampling on a denser grid or pattern.

The *marine geoid* is the term used to identify the seaward portion of the geoid. Mapping of the marine geoid is conducted by satellite altimetry. A radar sensor contained in a satellite platform sends radar pulses down to the ocean surface. A measurement of the distance between the satellite and sea surface topography can be made by measuring the round-trip travel time of the radar signal. Seeber, 1993 gives this distance as:

$a = c \dfrac{\Delta t}{2}$ where c is the speed of electromagnetic energy in a vacuum.

The sea surface topography or $\overline{H}$ is the deviation of the mean sea surface from the geoid. If mean sea level **is** understood to be a stationary sea surface corrected for time dependent variation, such a surface will vary from the geoid by 1 to 2 meters due to salinity differences, large-scale differences in atmospheric pressure, and strong ocean currents. The attempt to define the geoid by mean sea level only works to a precision of ±2 meters. The distance of the satellite platform from the level ellipsoid is the familiar ellipsoid height. Therefore: $h = N + \overline{H} + a$ where h is ellipsoid height, N is geoid height, $\overline{H}$ is the difference between a mean instantaneous sea level and the geoid, and (a) is the height observation from the satellite. Since most of the earth is ocean, the use of satellite altimetry since 1973 has contributed greatly to our knowledge of the geoid. Note that the equation above gives us, N, or geoid height. This result is given directly by satellite altimetry. The determination of geoid heights on land requires a different procedure.

Gravity Measurements

The force of gravity is measured at or near the surface of the earth in units of *milligals* (mgal). A milligal is a one millionth part of one "g", the unit used to measure the acceleration of military aircraft. One "g" is simply the accepted constant value for the acceleration of gravity, 9.81 m/s^2. Remember, gravity is the sum of gravitational acceleration and the centrifugal acceleration created by the earth's rotation. Gravitation is the stronger of the two accelerations and if a body is free to fall, it will accelerate toward the earth's center of mass. Two types of gravity measurements are possible: absolute gravity measurements and relative gravity measurements. Absolute gravity measurements determine the acceleration of gravity directly from observations made at a point. Up to this point in history, absolute gravity measurements require sophisticated measurement equipment and procedure in laboratory conditions. The few absolute gravity control stations that exist (approximately 50 in the United States) are part of the gravity datum — the International Gravity Standardization Net 1971 (IGSN71). The IGSN71 consists of 1854 gravity stations worldwide. The IGSN71 replaced the Potsdam Gravity Reference System founded upon the first absolute gravity station measured at Potsdam, Germany in 1906 with pendulum-type equipment.

The acceleration of gravity (g) can be computed from the measurement of the period (T) of a pendulum swung in a vacuum. The motion of the pendulum occurs in a vertical plane and is driven by the force of gravity. The period of a simple pendulum is a function of the length of the string (L) and gravity:

$$T = 2\pi \sqrt{\frac{L}{g}}$$

Therefore, if L is fixed in the construction of the device and T is measured, then g can be calculated. Pendulums have numerous associated error sources and are not sufficiently accurate to meet the needs of geodetic gravimetry.

In the United States, the basic falling body device (Hammond-Faller Apparatus) was jointly developed in the 1970's by J.A. Hammond of the Air Force Geophysics Laboratory and J.E. Faller of the Joint Institute for Laboratory Astrophysics. The Hammond-Faller Apparatus develops the acceleration of gravity by letting a corner cube fall in a vacuum. While the corner cube is falling, distance and time are measured by a laser beam.

Ballistic apparatus utilizing measurements on the rise and fall of a body have been developed to measure absolute gravity at the International Bureau of Weights and Measures at Sevres, France. At the beginning of the 21st Century, it can be said that absolute gravity measurements have been made at only a few key sites. The scarcity of absolute gravity measurements is due to the bulky and costly equipment in use. A single measurement takes days of careful work under laboratory conditions.

Relative gravity measurements have been used to densify the gravity network. Pendulum-type equipment was used up to the late 1960's. This equipment has been replaced by the modern gravimeter — the heart of which is a weight suspended on a sensitive spring. Changes in the length of the spring as the gravimeter is moved from place to place are recorded on a dial. These dial readings are proportional to gravity differences. Each individual instrument is empirically calibrated and this unique calibration factor is used to translate dial readings to gravity values.

Sound relative gravity survey practice and procedure is necessary to obtain precise gravity measurements. All gravity **surveys** require closure on known points to control calibration issues and drift. Drift is caused by instrument instabilities that cause the dial readings to change slowly with time unrelated to changes in the acceleration of **gravity.**

Gravity measurements provide values for the **acceleration** of gravity at points located on the surface of the earth. These gravity **measurements** must be converted to gravity anomalies. A gravity anomaly is the difference between a gravity observation reduced to sea level and normal gravity.

We have already mentioned the model of normal gravity in Chapter Eight in connection with the level ellipsoid. Normal gravity is the theoretical value representing the acceleration of gravity that would be generated by a uniform ellipsoidal earth. Therefore, the normal gravity at any point on the ellipsoid surface can be generated by formula. Torge, 1991 gives the normal gravity formula for the GRS 80 level ellipsoid as:

$$\gamma_o = 9.780327(1 = 0.0053024\sin^2\phi - 0.0000058\sin^2 2\phi)\,m/_{s^2}$$

Inspection of the formula indicates that normal gravity is a function of geodetic latitude.

Reduction of Gravity Measurements

A *gravity anomaly* is the difference between the actual (observed) acceleration of gravity at a point on the earth's surface reduced to the geoid surface and the computed normal acceleration of gravity of that same point on the level ellipsoid. A number of different procedures exist to reduce gravity observations to gravity anomalies. Each procedure will give different types of gravity anomalies. We shall consider two different anomalies: Free-air and Bouguer (pronounced "boo gay").

Free-Air Gravity Anomalies

The free-air reduction is so designated because it takes into account only the elevation of the station not the mass of the topography between the station and the geoid. Torge, 1991 gives the free-air anomaly as:

$$\Delta gf = g + \delta gf - \gamma_o$$

In the formula above, g is the surface gravity measurement, δgf is the partial derivative of gravity with respect to orthometric height, and γ_o is the normal gravity at the point on the level ellipsoid. Torge indicates that with respect to δgf :

$$\delta gf = \frac{\partial g}{\partial H} H \text{ is often approximated by } \frac{\partial \gamma}{\partial h} H .$$

The effect of the free-air reduction is as if the gravimeter were in free air at H meters above the geoid. While the free-air reduction neglects the influence of topography, the mass of the topography between the geoid and the observation point does exert an influence on the gravity measurement.

Bouguer Gravity Anomaly

The advantage of the Bouguer anomaly is to allow a topographic reduction. The Bouguer anomaly is stated as:

$$\Delta g_B = g - \delta g_{Top} + \delta gf - \gamma_o$$

162

Several methods exist for the evaluation of δg_{Top}. An estimate of the density (ρ) of the volume elements, that make up the topographic mass, gives the mass element dm = ρdV. A direct approach is to integrate or sum the influence of the mass elements (dm) each a distance (d) away from the surface point using the fundamental relationship:

$$dg = \frac{Gdm}{d^2}.$$

The sum of the gravitational influence provided by the mass elements is subtracted from the gravity measurement at the surface point. A problem with the Bouguer reduction is the determination of how large a volume about the surface point to include in the reduction. In addition, the Bouguer reduction assumes level terrain about the surface point. A terrain correction can be considered for irregularities in terrain about the surface station.

Geoid Heights

The distance from the ellipsoid to the geoid, measured along a normal to the ellipsoid, is known as the *geoid height, geoid separation,* or *geoid undulation.* Figure 9.2 shows the relationship between geoid height (*N*), ellipsoid height (*h*) and orthometric height (*H*) which may be expressed as: $h = H + N$. Within the

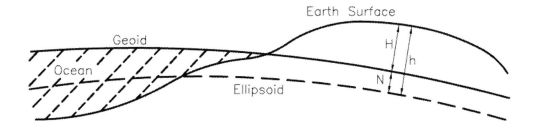

Figure 9.2 Geoid height (N), ellipsoid height (h), and Orthometric Height (H).

coterminous U.S. the geoid is located approximately 30 meters below the ellipsoid which results in negative values for geoid heights. The relationship between the geoid and ellipsoid in the continental U.S. calls for a different sketch shown in figure 9.3.

A British scientist, Sir George Gabriel Stokes, published a method by which the undulations of the geoid could be gained from gravity observations. Stokes function is given as:

$$S(\psi) = \csc\frac{1}{2}\psi + 1 - 5\cos\psi - 6\sin\frac{1}{2}\psi - 3\cos\psi \ln(\sin\frac{1}{2}\psi + \sin^2\frac{1}{2}\psi)$$

The angle (ψ) is the spherical distance between the point at which we wish to calculate N and the surface element (dσ) of the geoid where we know the mean gravity anomaly (Δg). The application of Stokes Theorem gives:

$$N = \frac{R}{4\pi\gamma} \iint_\sigma S(\psi)\Delta g\, d\sigma$$

Note that the theorem is not valid if masses exist outside the geoid. This explains the necessity for the reduction of gravity measurements to gravity anomalies. Secondly, the theorem provides the distance (N) from the geoid to the level ellipsoid which has the same volume as the geoid and the same center of mass. This implies that the theorem cannot be used to find the semimajor axis (a) of the level ellipsoid. We still have to depend on geometric geodesy for that determination.

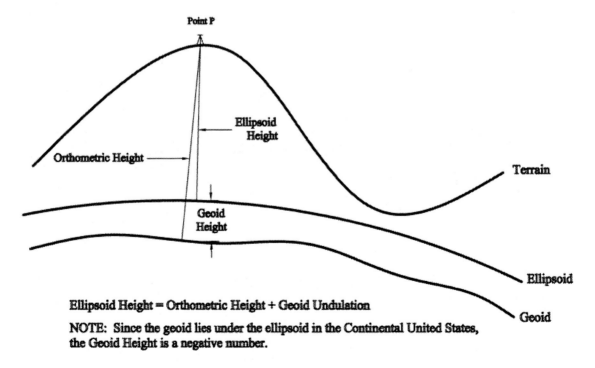

Ellipsoid Height = Orthometric Height + Geoid Undulation

NOTE: Since the geoid lies under the ellipsoid in the Continental United States, the Geoid Height is a negative number.

Figure 9.3 Relationship between geoid and level ellipsoid in continental U.S.

Torge, 1991 points out that in practice that mathematical integration is replaced by the summation of finite surface elements. Mean gravity anomalies (Δg) are determined for each surface grid element. Remember that since the geoid is an irregular surface, it must be measured point by point. It is logical to order this finite sampling into gridded mean gravity anomalies.

GEOID MODELS

Since a dense grid of gravity measurements does not exist over the United States, nor the world, the geoid is modeled by the information that does exist at any one point in time. The National Geodetic Survey has produced GEOID90, GEOID93, GEOID96 and the latest model of the geoid, *GEOID99*. Each successive model is characterized by more gravity measurements, associated data, and improvements in modeling.

The geoid is modeled by three components: the "long-wavelength" *global geopotential* component, the "medium-wavelength" *regional gravity anomaly* component, and the "short-wavelength" *local topography* component. Each component carries with it some degree of inaccuracy. The global geopotential component has historically been modeled at The Ohio State University (OSU). This model is produced from gravity readings taken around the world -- on land, at sea, within the atmosphere, and extra-terrestrial observations. The most recent global geopotential model is the *EGM 96* model jointly computed by the Goddard Space Flight Center and the National Imagery and Mapping Agency.

The regional gravity anomaly component is computed from gravity anomaly readings taken across an area. The gravity anomalies, derived from field observation, are used to refine the global geopotential model. The greater the number of anomalies, the greater the accuracy of the geoid model.

Local variations in topography have the least affect on the geoid model, but also offer the highest degree of refinement. Terrain variations are entered into the geoid model as a *digital elevation model* (DEM), typically input as a regularly spaced grid of spot elevations. The accuracy of the geoid model increases with decreasing spacing of the grid.

G99SSS

The G99SSS geoid model, computed by the NGS, is a purely gravimetric, geocentric geoid model covering the conterminous United States. The EGM 96 model was the basis for the *G99SSS* geoid model. Input data for the G99SSS consisted of 2.6 million terrestrial, ship, and altimetric gravity measurements, 30 arc second DEM data for most covered regions, one arc second DEM data for the Northwest USA (NGSDEM99), and the EGM96 global geopotential model. G99SSS was computed on a 1 arc minute by 1 arc minute grid covering the conterminous U.S.

GEOID99

GEOID99 is the current geoid model for the United States. It is described as a hybrid geoid model because it combines the G99SSS geoid model with datum transformations and NAD 83 GPS ellipsoid heights on NAVD 88 leveled benchmarks. GEOID99 incorporated 6169 GPS observed heights on NAVD 88 leveled benchmarks, roughly triple the number used in GEOID96. This provides more computed geoid heights to compare with modeled geoid heights. The least squares collocation model was used to identify a trend and signal in the geoid model output that was removed to provide an estimated 16% improvement over the output of GEOID96. Comparisons of GEOID99 computed heights with GPS determined geoid heights on incorporated benchmarks yields a 4.6 cm RMS difference. This value is used by NGS for the one sigma absolute accuracy of the GEOID99 model. This value is only valid in regions of GPS on NAVD 88 benchmark coverage (i.e., newer HARNS). NGS estimates a relative accuracy of approximately 2 ppm or better for all regions.

Input for GEOID99 consists of a NAD 83 latitude and longitude. Do not use NAD 27 positions for input.

Problem 9.1: Station KFALLS has NAD 83 geodetic coordinates 121°45'00.00000"W, 42°15'00.00000"N, 1300.000 m. Estimate the NAVD 88 orthometric height for BELL.

Output from GEOID99:

	latitude	longitude	N
Station Name	ddd mm ss.sssss	ddd mm ss.sssss	meters
K Falls	42 15 0.00000	121 45 0.00000 -	-22.509

$h_{KFALLS} = H_{KFALLS} + N_{KFALLS}$

$\therefore H_{KFALLS} = h_{KFALLS} - N_{KFALLS} = 1300.000 \text{ m} - (-22.509 \text{ m}) = 1322.509 \text{ m}$

Considering the stated absolute accuracy of GEOID 99, $H_{KFALLS} = 1322.509 \text{ m} \pm 0.046 \text{ m}$

GPS Leveling

Traditional differential leveling is capable of producing high accuracy orthometric height differences. Unfortunately, differential leveling is a relatively slow, painstaking process which translates into high costs. GPS allows surveyors to transfer geodetic control rapidly and at relatively low cost. It would be very desirable to take advantage of the speed of GPS if accurate orthometric height differences could be obtained by such a *GPS leveling* procedure.

GPS is capable of providing extremely accurate 3-D positions (approximately 1 cm + 1 ppm) with respect to the ellipsoid when operated in a *relative* or *differential* manner with two or more receivers. When a single GPS receiver is operated in a *point* or *absolute* positioning mode GPS provides positions (ϕ, λ, h) accurate from ±5 to 100 meters. This is quite unsuitable for most surveying applications. Thus, acceptable surveying accuracy of GPS is realized when a GPS vector of coordinate differences ($\Delta\phi$, $\Delta\lambda$, Δh or Δx, Δy, Δz) is determined from the relative positioning technique.

But even relative GPS provides accuracy with respect to the ellipsoid only, not with respect to the geoid. In order to determine orthometric heights from GPS, we must also know the geoid height as indicated by the formula: $H = h - N$. GEOID99 will provide the geoid height (N) at a point but the accuracy (±4.6 cm) isn't acceptable for even the least rigorous differential leveling standards (3rd Order: ±1.2√K mm).

The key to performing GPS leveling is to make use of the higher accuracy afforded by the relative technique. ΔH is the difference in orthometric heights between two points ($H_2 - H_1$). Δh is the ellipsoid height difference between two points provided by GPS (h_2 - h_1). This value has an estimated accuracy of ±(1 cm + 1 ppm) at the one sigma confidence level. The geoid height difference ΔN between two points (N_2 - N_1) may be determined from GEOID99 and has a predicted accuracy of ±(1 to 2 ppm). Since 1 ppm is 1 mm/km, the accuracy of GPS leveling is apparently limited by GPS and not the geoid model.

Field tests performed by the NGS using single-tie differential GPS for the determination of orthometric heights in conjunction with GEOID99 provide some insights into accuracies. Within reobserved HARN states (WA, OR, WI), the accuracy of GPS leveling over five and ten km baselines was approximately ±5 cm. This translates into a relative accuracy of 10 ppm for the shorter lines and 5 ppm for the longer lines. Second order, class II differential leveling over the same distances would yield relative accuracies of 0.6 ppm and 0.4 ppm, respectively.

A few words of warning regarding GPS leveling: do not perform GPS leveling with NGVD 29 elevations, use only NAVD 88 orthometric heights. Also, do not assume that you will be able to attain the accuracy values stated here without first performing some field tests! Use multiple points of known orthometric heights (vertical control) when performing GPS leveling, don't use a single benchmark as it will provide no check on your work. Project areas containing varied topography will typically produce poorer accuracies than flat terrain.

Problem 9.2: The following data is to be used to determine the orthometric height of station GPSBM.

H_{BM17} = 1277.234 m (NAVD 88)
$\Delta h_{BM17 \rightarrow GPSBM}$ = 134.257 m (NAD 83) per GPS observed baseline
N_{BM17} = -24.853 m (GEOID99)
N_{GPSBM} = -24.734 m (GEOID99)

$\Delta H = \Delta h - \Delta N$
$H_{GPSBM} - H_{KFALLS} = (h_{GPSBM} - h_{KFALLS}) - (N_{GPSBM} - N_{KFALLS})$
$H_{GPSBM} = H_{KFALLS} + \Delta h_{BM17 \rightarrow GPSBM} - (N_{GPSBM} - N_{KFALLS})$
H_{GPSBM} = 1277.234 m + 134.257 m – [-24.734 – (-24.853)]
H_{GPSBM} = 1411.372 m (NAVD 88)

Based upon the experience of the NGS, the accuracy of this value is likely to be on the order of ± 5-10 cm.

Study Questions

1. The distance between two first order horizontal control stations is computed by inverse to be 3,718.34 feet. The orthometric height at both stations is approximately 30 meters. The value of the geoid height at each station is negative and equal to the respective orthometric height. Given the relative positional tolerance for first order horizontal stations of 1/100,000, compute the allowable error expected in a field measurement between the two monuments.

2. Consider point A located on the terrain surface:
   a) Show the difference between an ellipsoid height and an orthometric height on a sketch at point A
   b) Under what conditions would the ellipsoid height equal the orthometric height at point A?

3. When would a geodetic distance (measured on the ellipsoid surface) between two points be longer than the corresponding horizontal distance on the terrain surface? Show your answer on a sketch.

4. If you found a bench mark on the surface of the geoid, and the geoid height was determined to be –29.451 meters at that place; what would be its:
   a) Orthometric height?
   b) Ellipsoid height?

5. If a surveyor knows the ellipsoid height at a place and adds the geoid undulation, what does the sum represent?

6. If the geoid height (N) at a place is negative, is the level ellipsoid above or below the geoid?

7. Given the following tables of data, estimate the orthometric height of station C.

Station	NAD 83 Position		NAVD 88 Orthometric Height
	Latitude	Longitude	
A	46°00'00.0000" N	113°00'00.0000" W	1200.000 m
B	46°05'00.0000" N	113°05'00.0000" W	1250.000 m
C	46°05'00.0000" N	113°00'00.0000" W	?

GPS Observations (NAD 83)		
From	To	Δh
A	C	18.457 m
B	C	-31.532 m

CHAPTER TEN

REDUCTION OF OBSERVATIONS

Observations refer to well-defined physical reference elements on the earth. A typical example is astronomic latitude, longitude, and azimuth measured with a theodolite from star observations. These measurements refer to the instantaneous rotation axis, the instantaneous equator of the earth, and the local astronomic horizon, which is the plane perpendicular to the local plumb line. Vertical angle and horizontal angle observations are referenced with respect to the plumb line and the local astronomic horizon. Even though astronomic latitude, longitude, and azimuth determinations are less frequent today than in the pre-satellite era, these quantities are still conceptually important when defining the geodetic reference frame as a basis for processing all observations in surveying. Leick, 1995.

The Plumb Line

A *plumb bob* consists of a length of string (or wire) attached to a small pointed mass of metal (the bob). When the plumb bob is suspended, the string indicates the average direction of gravity (the plumb line) between the point of suspension and the bob. The plumb line is perpendicular to each level surface it intersects. Chapter Four introduced the concept of the level or equipotential surface. The geoid is just one of many equipotential surfaces. While the difference in potential is constant between two level surfaces, the distance between these level surfaces is not constant. In other words, equipotential surfaces are not parallel. Since the plumb line is perpendicular to each level surface, the plumb line cannot be a straight line. In fact the plumb line is best described as a space curve (a curve in three dimensions). A significant phenomenon of our physical world, that has an effect on survey measurements, is the fact that the plumb line is not a straight line, but a space curve.

The second physical phenomenon concerning the plumb line is that it is pulled or deviates toward concentrations in the earth's mass. For example, mountains represent a mass concentration and the oceans a mass deficiency. Generally, the plumb line is pulled toward the mountains and away from the seas. The effects of the uneven distribution of mass in the earth's physical construction are termed *mass anomalies*. These mass anomalies account for the deviation of the plumb line from the geometrically determined normal to the ellipsoid.

172

The level bubble on a survey instrument is used to bring the first (vertical) axis of the instrument coincident with the plumb line. The local horizon (tangent) plane is perpendicular to the plumb line at the instrument. Since the mass of the earth is not uniform, the direction of the plumb line determined by gravity deviates from the normal to the ellipsoid at any point on the earth's surface. Therefore, an observation or measurement made by a survey instrument is classified as an astronomic observation. To accomplish precise geodetic calculations, these astronomic observations must be converted to their corresponding geodetic values through a process known as *reduction of observations*. Common astronomic observations are astronomic latitude (Φ), astronomic longitude (Λ), and astronomic azimuth (A). The conversion from astronomic to geodetic can be accomplished by understanding the concept of the 'deflection of the vertical.'

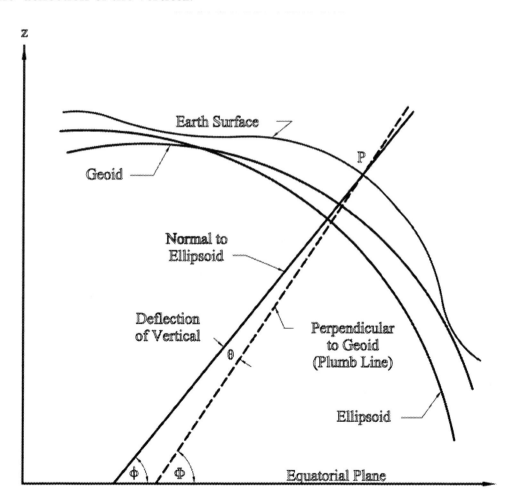

Figure 10.1 Deflection of the Vertical.

173

Deflection of the Vertical

The lack of parallelism between the geoid and the ellipsoid and the curvature of the plumb line results in a lack of coincidence between the plumb line and a normal to the ellipsoid at any point on the earth's surface. The angle between these two lines is known as the *deflection of the vertical*. The deflection of the vertical is illustrated in figure 10.1. The deflection of the vertical is measured from the ellipsoidal normal to the plumb line and may be resolved into a north south component, ξ (xi), and an east west component, η (eta). If the astronomic azimuth, A, of the deflection (θ) is known, the components may be computed from: $\xi = \theta \cos A$ and $\eta = \theta \sin A$.

The deflection of the vertical may be determined by two methods: 1) at the geoid surface using a contour map of geoid heights; or 2) at the earth surface using the *DEFLEC99* model produced by the National Geodetic Survey (NGS). The first method is illustrated in figures 10.2 and 10.3 and may be computed from:

$$\theta = \arctan\left(\frac{\Delta N}{S}\right)$$

ΔN is the change in geoid height and S is the geodetic distance. The astronomic azimuth of the deflection of the vertical is estimated from the direction of a line perpendicular to the geoid height contours.

Problem 10.1: Determine the deflection of the vertical, θ, and its components, ξ and η, for the scenario shown in figures 10.2 and 10.3.

The geoid height contours shown in figure 10.2 represent the deviation of the geoid from the ellipsoid. The angle of this deviation is equal to the deflection of the vertical as shown in figure 10.3.

$$\theta = \arctan\left(\frac{\Delta N}{S}\right) = \arctan\left(\frac{-21.8\ m - (-22.1\ m)}{6370\ m}\right) = 0°00'09.7"$$

$\xi = \theta \cos A = 9.7" \cos 325° = 8.0"\ (North);$ $\eta = \theta \sin A = 9.7" \sin 325° = -5.6"\ (West).$

$\therefore$ The plumb line lies northwest of the ellipsoidal normal.

DEFLEC99

The NGS DEFLEC99 model for deflection of the vertical is based upon the GEOID99 model. It is important to realize that the deflection of the vertical values are derived from the shape of the geoid. The shape of the geoid is modeled by GEOID99. Deflection of the vertical values are first computed using the geoid slopes on the GEOID99 model. The values are then extended up to the surface of the earth (the topography) by applying a correction for the curvature of the plumb line.

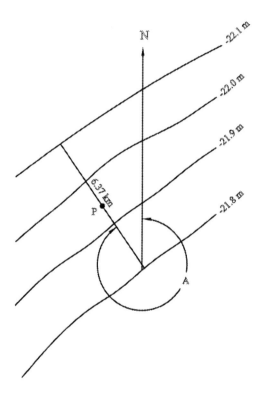

Figure 10.2 Contour map of geoid heights.

DEFLEC99 provides earth surface values for the deflection of the vertical in the continental U.S., Hawaii, Puerto Rico, and the U.S Virgin Islands. A deflection interpolation program and deflection files for the personal computer can be downloaded from the National Geodetic Survey web site, however the output can be gained directly over the Internet from interactive use of the Geodetic Tool Kit. The National Geodetic Survey designates the component in the meridian plane as "Xi" and the component in the prime vertical as "Eta."

The components of the deflection of the vertical for the main campus of Troy State University (TSU) in Troy, Alabama are: Xi = 0.14 arc seconds and Eta = -2.42 arc seconds. A positive Xi indicates that the astronomic latitude will fall to the north of the geodetic latitude at this station. A negative Eta means the astronomic longitude will fall to the west of the geodetic longitude at this station.

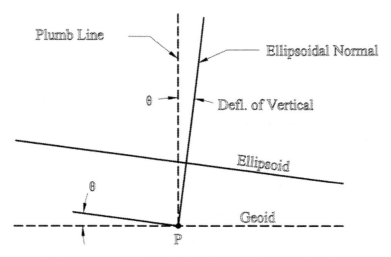

Figure 10.3 Determining deflection of the vertical.

The DEFLEC99 program also gives the computed Laplace correction termed "Hor. Laplace." The Hor. Laplace for the station at Troy State University is given by DEFLEC99 as 1.49 arc seconds. The value is added to an astronomic azimuth to gain a near-geodetic azimuth. The reference to 'near-geodetic' is made because the adjustment 'Hor. Laplace' consists of the term $(-\eta \tan \phi)$ and does not contain the deflection correction needed for inclined lines of sight. The deflection correction is:

$$[\xi \sin \alpha - \eta \cos \alpha] \cot(z\angle)$$

$z\angle$ is the zenith distance or angle. Note that use of the deflection correction is not necessary for horizontal lines of sight because the magnitude of the correction becomes negligible. Further discussion of the deflection correction is found in a later section within this chapter dealing with *Reduction of Observed Horizontal Angles*.

Although the magnitude of the deflection of the vertical is typically just a few seconds, magnitudes can approach an arc minute in the Rocky Mountain

176

States and in the U.S. Virgin Islands. Significant deflections are also found along the Pacific coast where mountains meet the ocean.

Problem 10.2: Determine the earth surface deflection of the vertical for station KFALLS having NAD 83 geodetic coordinates 42°15'00.00000"N, 121°45'00.00000"W. Output from DEFLEC99:

Station Name	latitude ddd mm ss.sssss	longitude ddd mm ss.sssss
K Falls	42 15 0.00000	121 45 0.00000

Xi arc-sec	Eta arc-sec	Hor Lap arc-sec
-5.67	-5.00	4.54

Definition of Astronomic Observations

Several procedures involve making measurements on heavenly bodies to determine the latitude and/or longitude of a place or the azimuth of a line. The fact that the survey instrument is leveled up to the plumb line defines the computed result of these measurements as astronomic observations.

The astronomic latitude (Φ) at a place is the angle between the plumb line at that place and the plane of the earth's equator. Astronomic longitude (Λ) is defined exactly the same way as geodetic longitude. It differs from geodetic longitude by the longitudinal component of the deflection of the vertical. Astronomic azimuth (A) of a line is the horizontal angle from the astronomic-determined meridian measured clockwise from north.

Geodetic - Astronomic Conversions

One of the uses of the deflection of the vertical is for performing conversions between geodetic values (latitude, longitude and azimuths) and astronomic values. Geodetic latitude, longitude and azimuth are reckoned with respect to the ellipsoid. Astronomic latitude, longitude and azimuth are determined by celestial observations made using optical surveying instruments (theodolites, total stations, etc.). Since surveying instruments are oriented with respect to gravity (the vertical axis is coincident with the plumb line) and not the ellipsoid, the deflection of the vertical may be used to convert observed (astronomic) values to their geodetic counterparts.

The following formulas are useful in performing geodetic - astronomic conversions:

$$\phi = \Phi - \xi$$

$$\lambda = \Lambda - \frac{\eta}{\cos \phi}$$

$$\alpha = A - \eta \tan \phi \quad or \quad \alpha = A + Hor.\ Laplace$$

Note the two forms available for conversion between geodetic azimuths (α) and astronomic azimuths (*A*). The term ($\eta \tan\phi$) is known as the *Laplace correction* and is subtracted from the astronomic azimuth to yield the geodetic azimuth. This is seen the following formula:

$$\alpha = A - \eta \tan \phi$$

The *Hor. Laplace* correction is from *DEFLEC99* and is equal to ($-\eta \tan\phi$) so it must be added to gain the geodetic azimuth

Problem 10.3: The following data was obtained by celestial observations at station KFALLS: Φ_{KFALLS} = 42°14'56.7"N, Λ_{KFALLS} = 121°45'10.3"W, $A_{KFALLS\rightarrow BIG}$ = 60°00'00.0". Compute the geodetic position of KFALLS and the geodetic azimuth from station KFALLS to station BIG. Use values from Problem 10.2.

$$\phi_{KFALLS} = \Phi_{KFALLS} - \xi_{KFALLS} = 42°14'56.7" - (-5.67") = 42°15'02.4"\ N$$

$$\lambda_{KFALLS} = \Lambda_{KFALLS} - \frac{\eta_{KFALLS}}{\cos\phi_{KFALLS}} = -121°45'10.3" - \frac{-5.00"}{\cos 42°15'02.4"}$$

$$\lambda_{KFALLS} = -121°45'03.5" = 121°45'03.5"W$$

$$\alpha_{KFALLS\rightarrow BIG} = A_{KFALLS\rightarrow BIG} + Hor.Lap._{KFALLS} = 60°00'00.0" + 4.54" = 60°00'04.5"$$

The concepts that we have examined to this point may appear to have little practical application. However, contemporary surveyors, unlike their predecessors, are now often required to integrate classical (terrestrial) observations with GPS determined positions. The mechanics of accomplishing this task may seem obscure but it will be seen that application of the fundamental concepts presented thus far allow this integration to be handled systematically.

"Horizontal Distance" - What Is It?

By now you are probably convinced that there is a bit more to surveying than arbitrarily assigning coordinates to a point (say 10,000N, 10,000E), assuming a bearing to a backsight (North seems good!), and considering the earth as a plane. Since we are in fact working on a curved surface, what do we consider as being "horizontal distance"? The "flat earth" assumption is shown in figure 10.4, the plumb lines are assumed to be parallel and right angle trigonometry is used to reduce observed slope distance to "horizontal distance".

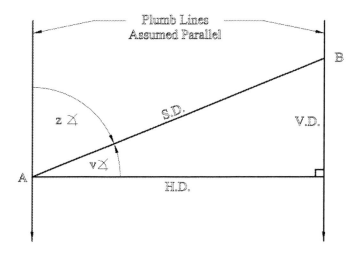

Figure 10.4 Flat earth assumption.

Under what conditions is the simple right angle reduction of slope distance to horizontal distance no longer a suitable assumption? The magnitude of the distance and the change in orthometric height between two stations dictate the degree of discrepancy between the right angle (parallel plumb lines) assumption and reality. Figure 10.5 will be used to illustrate this point. We will assume that the earth is a sphere having a radius of 6,371 km. Assume the distance between points A and B is 1000 meters measured along the surface of the sphere.

Now assume the distance is measured using an *electronic distance measuring instrument* (EDMI) and prism which are two meters above the surface of the sphere. The loss of accuracy (s) by assuming parallel plumb lines is negligible and may be computed from:

$$\theta = \frac{1000\,meters}{6,371,000\,meters} = 0.000157\,radians = 0°00'32"$$

$$s = 2\,meters \times 0.000157\,radians = 0.0003\,meter$$

179

Now clearly 0.3 mm cannot be perceived when a typical EDMI has a random error component of approximately ±(2 mm + 2 ppm). But suppose the change in elevation from A to B is significant and B is actually located at B' that is 200 m higher than B. Now a failure to recognize that the plumb lines converge results in appreciable error:

$$s' = 202\ meters \times 0.000157\ radians = 0.032\ meters$$

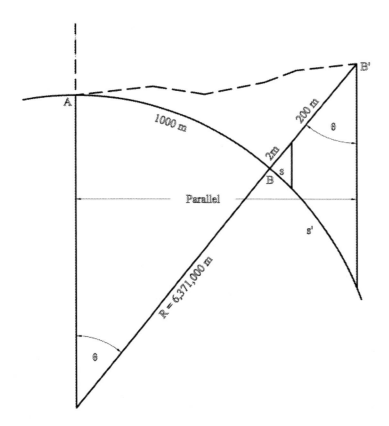

Figure 10.5 Example of plumb line convergence.

The prudent surveyor should not ignore the fact that the earth is curved! But still we have the dilemma: What is "horizontal distance"? Figure 10.6 illustrates several different types of distance between points A and B on the earth surface.

Distance D_1 is the slope distance observed from the EDMI to the prism. The mark-to-mark slant range is denoted D_2. Distance D_3 is a near-level distance measured along the elevated sphere (ellipsoid) surface at an ellipsoid height of h_A while D_4 is the chord of that arc. The geodetic distance is denoted D_5 and measured along the ellipsoid surface.

Distances D_3 and D_4 are identical for all practical purposes (within 0.001 m in 10 km) and are useful in traversing with state plane coordinates. This distance will be referred to as *"level distance"* or LD. The computation of D_4 is presented below.

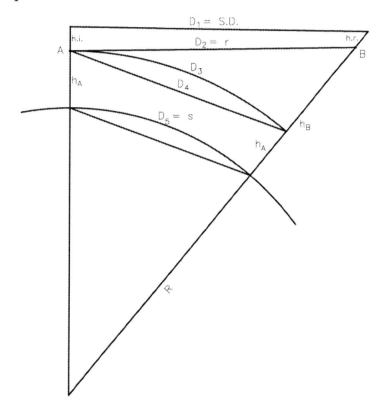

Figure 10.6 Different "horizontal distances."

Level Distance

The level distance computed between the plumb lines at the instrument's ellipsoid height is necessary in order to reduce an observed slope distance (D_1) to geodetic distance (D_5 or S). The geodetic distance is needed when performing state plane coordinate computations. Geodetic distance may also be required to perform traversing with geodetic coordinates.

Most total stations will compute an approximate level distance when atmospheric refraction and earth curvature corrections are applied to observed slope distances for internal computation of horizontal distance. These

181

corrections are correct for distance measured on the ellipsoid. They do not, however, account for measurements observed from stations located above (or below) the ellipsoid. The formulas presented below account for ellipsoid height. These formulas are made with reference to figure 10.7.

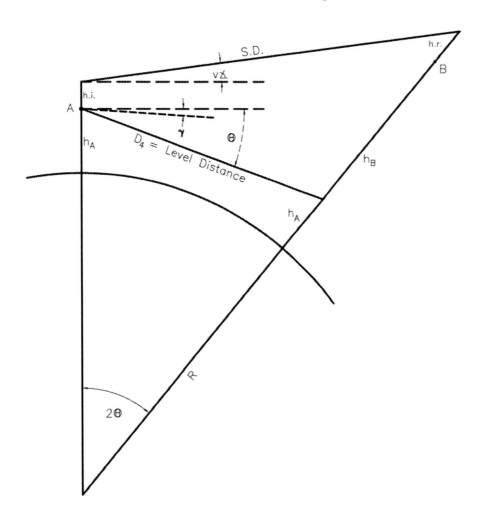

Figure 10.7 Computation of D_4 – "level distance."

Correction for Atmospheric Refraction

Atmospheric refraction is the term given to the phenomenon of light bending through the atmosphere. This causes observed zenith angles and vertical angles to be incorrect since they are actually measured to the tangent to the light arc at the instrument. This can be overcome by observing *reciprocal* angles, that is, observing the zenith or vertical angle from both ends of the line simultaneously. The correct zenith angle is obtained by taking the mean of the endpoint observations.

182

Oftentimes, the observation of reciprocal zenith angles is impractical for the typical surveyor and the angle is only observed from one end of the line. When this is the case, the observed zenith angle must be corrected for atmospheric refraction. The corrected zenith angle is found from:

$$z\angle_{corr.} = z\angle_{obs.} + \gamma = z\angle_{obs.} + k\left[\frac{S.D.\sin(z\angle_{obs.})}{R+h_A}\right]\left(\frac{180°}{\pi}\right)$$

where $z\angle_{corr.}$ is the zenith angle that has been corrected for atmospheric refraction, $z\angle_{obs.}$ is the observed zenith angle, k is the coefficient of atmospheric refraction, R is the mean earth radius, and h_A is the ellipsoid height of the instrument station. A value of 0.14 is typically used for (k) although it is affected by humidity, temperature and barometric pressure.

Correction for Earth Curvature

The curvature of the earth correction would be more accurately described as the "plumb line convergence correction", but never seems to be referred to as such. Regardless, this correction accounts for the fact that the plumb lines are not parallel at different locations on the earth'' surface. The correction, x, is estimated using a spherical earth assumption:

x = S.D. sinδ tanθ

where δ = θ + (90° - $z\angle_{corr.}$)

$$\theta = \frac{S.D.\sin(z\angle_{corr.})}{2(R+h_A)}\left(\frac{180°}{\pi}\right).$$

The level distance between points A and B may the be computed using

L.D. = S.D. cosδ - x.

Problem 10.4: The observed slope distance from station KFALLS to station WELL is 2497.176 m at an observed zenith angle of 93°05'15". The ellipsoid height of station KFALLS is 937.7 m. Compute the level distance from KFALLS to WELL.

First, apply atmospheric refraction correction to the observed zenith angle.

183

$$z\angle_{\text{corr.}} = z\angle_{\text{obs.}} + k\left[\frac{\text{S.D.}\sin(z\angle_{\text{obs.}})}{R + h_A}\right]\left(\frac{180°}{\pi}\right) = 93°05'15" + 0.14\left[\frac{2497.176\text{ m}(\sin 93°05'15")}{6,371,000\text{ m} + 937.7\text{ m}}\right]\left(\frac{180°}{\pi}\right)$$

$$z\angle_{\text{corr.}} = 93°05'15" + 0°00'11" = 93°05'26"$$

Compute the earth curvature correction

$$\theta = \frac{\text{S.D.}\sin(z\angle_{\text{corr.}})}{2\,(R + h_A)}\left(\frac{180°}{\pi}\right) = \frac{2497.176\text{ m}\,(\sin 93°05'26")}{2(6,371,000\text{ m} + 937.7\text{ m})}\left(\frac{180°}{\pi}\right) = 0°00'40"$$

$\delta = \theta + (90° - z\angle_{\text{corr.}}) = 0°00'40" + (90° - 93°05'26") = -3°04'46"$

$x = \text{S.D.}\sin\delta\tan\theta = 2497.176\text{ m }(\sin\text{-}3°04'46")\tan 0°00'40" = -0.026\text{ m}$

$\therefore \text{L.D.} = \text{S.D.}\cos\delta - x = 2497.176\text{ m }(\cos\text{-}3°04'46") - (\text{-}0.026\text{ m}) = 2493.596\text{ m}$

Geodetic Distance

Geodetic distance is measured on the surface of the ellipsoid. Assuming that the level distance has been computed, the geodetic distance may be readily computed from a simple proportion. Figure 10.6 illustrates the proportion based on radius. The following equation may be used to compute geodetic distance, S, from level distance, L.D. (= D_3):

$$S = \text{L.D.}\left(\frac{R}{R + h}\right).$$

Expressing ellipsoid height in terms of orthometric height, H, and geoid height, N, we may write

$$S = \text{L.D.}\left(\frac{R}{R + H + N}\right).$$

Problem 10.5: Compute the geodetic distance between stations KFALLS and WELL for the scenario described in Problem 10.4

$$S = 2493.596\text{ m}\left(\frac{6,371,000\text{ m}}{6,371,000\text{ m} + 937.7\text{ m}}\right) = 2493.229\text{ m}$$

Traversing with Geodetic Coordinates

There are many instances when it is impractical to use GPS for positioning in the field. Vegetative canopy may preclude the use of GPS or the surveyor may not own GPS equipment. When using conventional surveying instruments and traversing for geodetic positioning, the surveyor must be acutely aware of the need for using observations referred to the ellipsoid. As previously noted, observations made with conventional surveying instruments are with respect to the plumb line. It is not practical to orient the vertical axis of an instrument with the ellipsoidal normal at the occupied point. Alternatively, observed values are "reduced" to geodetic (ellipsoidal) values. Observed angles, both horizontal and zenith, must be reduced to geodetic values. These additional angle reductions are investigated in this section.

Reduction of Observed Zenith Angles

The zenith angle corrected for atmospheric refraction ($z\angle_{corr.}$) may be reduced to a geodetic zenith angle ($z\angle_{geod.}$) using $z\angle_{geod.} = z\angle_{corr.} + \xi \cos\alpha + \eta \sin\alpha$ where ξ and η are the components of the deflection of the vertical at the instrument and α is the geodetic azimuth from the instrument to the observed station.

Problem 10.6: Compute the geodetic zenith angle from station KFALLS to station WELL using the data given in Problems 10.2 and 10.4. Assume the geodetic azimuth from KFALLS to WELL is 200°00'00".

$$z\angle_{geod.} = z\angle_{corr.} + \xi \cos\alpha + \eta \sin\alpha$$
$$= 93°05'26" + (-5.67") \cos 200°00'00" + (-5.00") \sin 200°00'00"$$
$$= 93°05'33"$$

Reduction of Observed Horizontal Angles

The rigorous reduction of observed horizontal angles to geodetic horizontal angles is accomplished using the following formula $H\angle_{geod.} = H\angle_{obs.} + \Delta H\angle_1 + \Delta H\angle_2 + \Delta H\angle_3$ where $H\angle_{geod.}$ is the reduced geodetic horizontal angle, $H\angle_{obs.}$ is the observed horizontal angle, $\Delta H\angle_1$ is the correction for deflection of the vertical, $\Delta H\angle_2$ is the correction for skewness of the ellipsoidal normals, and $\Delta H\angle_3$ is the correction from the normal section to the geodesic. These last two corrections are negligible for most practical purposes (<0.5" combined). Thus the simplified reduction formula may be written $H\angle_{geod.} = H\angle_{obs.} + \Delta H\angle$ where $\Delta H\angle$ is the correction for deflection of the vertical. This correction is the result of the combined effect of the deflection of the vertical on sighting the backsight and the foresight. The correction may be computed from:

$$\Delta H\angle = \delta H\angle_{BS} - \delta H\angle_{FS} \quad \text{where:}$$

$$\delta H\angle_{BS} = -(\xi \sin\alpha_{BS} - \eta\cos\alpha_{BS})\cot(z\angle_{geod.(BS)})$$

$$\delta H\angle_{FS} = -(\xi \sin\alpha_{FS} - \eta\cos\alpha_{FS})\cot(z\angle_{geod.(FS)}).$$

The deflection of the vertical components (ξ, η) are measured at the instrument station.

Problem 10.7: A horizontal angle (to the right) of 230°00'00" is observed at station KFALLS when backsighting station WELL and foresighting station BARN. The zenith angle from KFALLS to BARN is 88°00'00". Compute the geodetic azimuth from KFALLS to BARN.

From Problem 10.6, $\alpha_{\text{KFALLS}\rightarrow\text{WELL}}$ = 200°00'00".

Compute the approximate azimuth to station BARN
$$\alpha_{\text{KFALLS}\rightarrow\text{BARN}} \approx \alpha_{\text{KFALLS}\rightarrow\text{WELL}} + H\angle_{obs.} \approx 200° + 230° - 360° \approx 70°$$

Compute the geodetic horizontal angle

$$\delta H\angle_{BS} = -(\xi\sin\alpha_{BS} - \eta\cos\alpha_{BS})\cot(z\angle_{geod.(BS)})$$
$$= -[(-5.67")\sin 200°00'00" - (-5.00")\cos 200°00'00"]\cot 93°05'26" = -0.15"$$

$$\delta H\angle_{FS} = -(\xi\sin\alpha_{FS} - \eta\cos\alpha_{FS})\cot(z\angle_{geod.(FS)})$$
$$= -[(-5.67")\sin 70° - (-5.00")\cos 70°]\cot 88°00'00" = 0.03"$$

$$\Delta H\angle = \delta H\angle_{BS} - \delta H\angle_{FS} = -0.15" - 0.03" = -0.18"$$

For most practical applications, this correction is insignificant. However, the correction should be computed and evaluated when performing precise traversing.

Reduction of Distances

When traversing with geodetic coordinates it is most efficient to incorporate local geodetic horizon coordinates as introduced in Chapter Seven. The formulas for e, n and u developed in Chapter Seven required use of the slant range. Rather than computing the slant range from the observed slope distance (a slightly laborious task!), the slope distance may be resolved into relevant components in the e-n plane and the up direction. Figure 10.8 shows a typical situation, viewed perpendicular to the vertical plane containing the observed slope distance. For the generalized case shown in the figure, the rod height (h.r.) is not equal to the instrument height (h.i.).

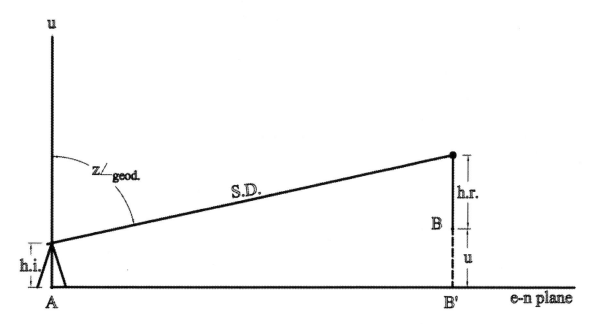

Figure 10.8 Slope distance reduction for local geodetic horizon coordinates.

The e-n plane component of the slope distance, AB' in figure 10.8, may be computed AB' = S.D. sin z∠$_{geod.}$. This horizon plane component may then be resolved into e and n components using the geodetic azimuth of line AB.

e = S.D. sin z∠$_{geod.}$ sin α

n = S.D. sin z∠$_{geod.}$ cos α

The u coordinate for point B, the perpendicular distance from the e-n plane, may be computed:

u = BB' = S.D. cos z∠$_{geod.}$ + hi - h.r.

The scenario shown in figure 10.8 assumes that the plumb lines are parallel at the instrument and the rod. In reality this assumption is not true. The observed slope distance actually needs to be shortened, slightly, by an amount equal to:

187

$$\frac{\text{S.D.(h.r.)}}{R + h_A}$$

Since this distance correction is insignificant, it will be ignored.

Problem 10.8: A total station instrument is set up at station KFALLS. Station BS is the backsight and station FS is the foresight. Given the field observations listed below, compute the geodetic coordinates of station FS.

Station Coordinates (NAD 83)			
Station	Latitude	Longitude	Ellipsoid Ht.
KFALLS	42°15'00.00000" N	121°45'00.00000" W	1300.000 m
BS	42°20'00.00000" N	121°50'00.00000" W	1350.000 m
FS	?	?	?

Observations to Station FS (Instrument at KFALLS)
Horizontal angle right = 150°00'00"
Zenith angle = 92°00'00"
h.i. = 1.521 m (at KFALLS)
h.r. = 2.184 m (at FS)

Deflection of the Vertical Components (at KFALLS, from Problem 10.2)
ξ = -5.67"
η = -5.00"

Step 1: Determine azimuth and zenith angle to BS
    This task may be most readily accomplished using the NGS *INVERS3D* program found at the NGS Web site. It is more practical, however, for the student to compute these values using local geodetic horizon coordinates with station KFALLS being the origin of the local system.

Local Geodetic Horizon Coordinates			
Station	e	n	u
KFALLS	0.000 m	0.000 m	0.000 m
BS	-6869.529 m	9261.898 m	39.569 m

$$\alpha_{KFALLS \to BS} = \arctan\left(\frac{e}{n}\right) = \arctan\left(\frac{-6869.529\ m}{9261.898\ m}\right) = 323°26'09"$$

$$z\angle_{geod \cdot KFALLS \to BS} = \arctan\left(\frac{\sqrt{e^2 + n^2}}{u}\right) = \arctan\left(\frac{\sqrt{(-6869.529\ m)^2 + (9261.898\ m)^2}}{39.569}\right)$$

$$= 89°48'12"$$

Step 2: Compute geodetic zenith angle to FS

$$z\angle_{corr.} = 92°00'00'' + 0.14\left[\frac{1500.000\ \text{m}(\sin 92°00'00'')}{6,371,000\ \text{m} + 1300\ \text{m}}\right]\left(\frac{180°}{\pi}\right) = 92°00'00'' + 0°00'07'' = 92°00'07''$$

$\alpha_{KFALLS \to FS} \approx \alpha_{KFALLS \to BS} + H\angle_{obs.} \approx 323°26'09'' + 150° - 360° \approx 113°26'09''$
$z\angle_{geod.} = 92°00'07'' + (-5.67'') \cos 113°26'09'' + (-5.00'') \sin 113°26'09''$
        $= 92°00'05''$.

Step 3: Compute the azimuth to FS

$$\delta H\angle_{BS} = -[(-5.67'')\sin 323°26'09'' - (-5.00'')\cos 323°26'09'']\cot 89°48'12'' = -0.03''$$

$$\delta H\angle_{FS} = -[(-5.67'')\sin 113°26'09'' - (-5.00'')\cos 113°26'09'']\cot 92°00'05'' = -0.25''$$

$$\Delta H\angle = \delta H\angle_{BS} - \delta H\angle_{FS} = -0.03'' - (-0.25'') = 0.22''$$

This correction is negligible given the implied precision of the observed horizontal angle (±0.5").

$\therefore \alpha_{KFALLS \to FS} = 113°26'09''$

Step 4: Compute LGH coordinates for FS

e = 1500.000 m (sin 92°00'05") sin 113°26'09" = 1375.420 m

n = 1500.000 m (sin 92°00'05") cos 113°26'09" = -596.219 m

u = 1500.000 m (cos 92°00'05") + 1.521 m – 2.184 m = -53.049

Step 5: Compute geocentric and geodetic coordinates for FS

x = -2,487,662.751 m	ϕ = 42°14'40.67619" N
y = -4,022,632.013 m	λ = 121°44'00.01714" W
z = 4,266,596.644 m	h = 1247.127 m

Line of Sight Obstructions

Geodetic horizon coordinates may be used to determine whether line of sight obstructions exist between points, particularly when the points are a considerable distance apart. Figure 10.9 shows a typical scenario. It is desired to have line of sight between points A and B. Point C represents a possible obstruction.

Assuming that geodetic coordinates have been obtained for all points (using GPS or other methods), local geodetic coordinates may be computed for all points using one of the endpoints (A or B) as the origin of the local system.

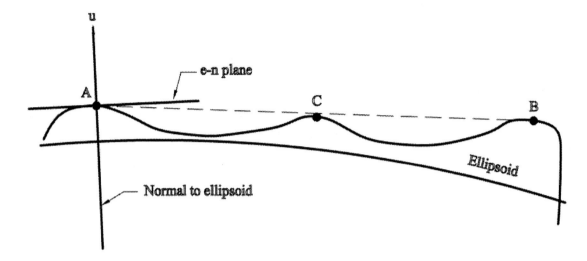

Figure 10.9 Possible line of sight obstruction between points A and B.

Whether point C represents a vertical obstruction to the line of sight between A and B may be determined using vertical angles as shown in figure 10.10. The vertical obstruction is calculated using simple right angle trigonometry from the slant range to C and the difference in vertical angles.

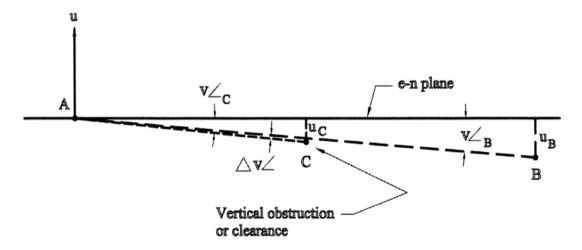

Figure 10.10 Calculation of vertical obstruction.

By inspection, if the vertical angle of C (v∠$_C$) is greater than (in an algebraic sense) the vertical angle of B (v∠$_B$), no obstruction will exist. The vertical obstruction (or clearance) may be computed from (r sin(Δv∠)) where r is the slant range to point C and Δv∠ is the difference in vertical angles.

Whether point C represents a horizontal obstruction is best determined using the azimuths from A to B (α$_B$) and C (α$_C$). The difference in azimuths (Δα) and the e-n plane distances allow computation of the horizontal obstruction or clearance (figure 10.11) from:

$$\sqrt{e_C^2 + n_C^2}\ \sin(\Delta\alpha)$$

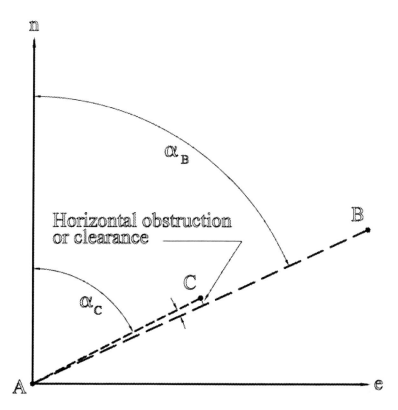

Figure 10.11 Computation of horizontal obstruction.

Problem 10.9: It has been proposed to construct cell phone towers at points A and B. Topographic maps indicate a potential line-of-sight obstruction between A and B at point C. GPS observations yield the following table of coordinates for the three points. Determine whether point C is an obstruction if 100 m of clearance is needed.

191

GPS Derived Coordinates (NAD 83)			
Point	x (m)	y (m)	z (m)
A	-2,588,082.624	-3,867,315.761	4,349,199.752
B	-2,723,863.561	-3,851,167.247	4,273,569.903
C	-2,673,111.729	-3,857,417.072	4,302,128.603

Using point A as the origin of the local geodetic horizon system, compute local coordinates for points B & C.

LGH Coordinates			
Point	e (m)	n (m)	u (m)
A	0.000	0.000	0.000
B	-121,824.731	-97,633.279	-6,602.215
C	-76,170.461	-61,050.789	-3,806.292

Compute angles and distances from A to B and C.

All Values from Point A				
Point	V∠	r (m)	α	e-n plane dist (m)
B	-2°25'17.6"		231°17'25.7"	
C	-2°13'58.6"	97,691.483	231°17'15.9"	97,617.304

$\Delta v\angle$ = -2°13'58.6" – (-2°25'17.6") = 0°11'19.0"
Vertical clearance = r sin($\Delta v\angle$) = 97,691.483 m (sin 0°11'19.0") = 321.588 m.

∴There is no obstruction. The horizontal clearance may be computed as follows:

$\Delta\alpha$ = 231°17'25.7" – 231°17'15.9" = 0°00'09.8"
Horizontal clearance = $\sqrt{e_C^2 + n_C^2}$ sin($\Delta\alpha$) = 4.638 m.

∴The total clearance = 321.621 m.

Study Questions

1. Determine the deflection of the vertical and its components at station TALL having NAD 83 coordinates (39°N, 106°W).
2. A slope distance of 13,877.56 US feet is observed from station TALL (see question 1) to station SHORT at a zenith angle of 85°. Determine the geodetic distance from TALL to SHORT if the NAVD 88 orthometric height of TALL is 6,781.54 US feet.
3. Your job is to determine the geodetic coordinates (NAD83-1998) of station END. You accomplish this by traversing from station SKILLET (PID AI2012). Station ROCKY (PID AI2004) is used as the initial backsight. Apply corrections to zenith angles! The summarized field notes are:

192

Inst. @ SKILLET, BS: ROCKY, FS: END
H∠ = 215°30'10"
SD = 1835.726 m
Z∠ = 88°20'30"
hi = 1.427 m
hr = 2.246 m

4. You are hired to determine if it is feasible to construct microwave relays in eastern Oregon. It is proposed to construct one relay on Diamond Peak and another on Lookout Mountain. Inspection of a topographic map of the state reveals that Maiden Peak may cause an obstruction to the line-of-sight requirement (100 meters) for proper operation of the relays. Your GPS survey results in the following coordinates (GRS 80) for the proposed relay locations:

Diamond Pk.	Lookout Mtn.
λ = 122°08'58.5"W	λ = 121°41'54.3"W
φ = 43°31'14.5"N	φ = 43°48'17.0"N
h = 2644.8 m	h = 1872.3 m

The potential obstruction of Maiden Peak causes you to make GPS observations on top of the mountain with the following result:

λ = 121°57'53.0" W; φ = 43°37'36.4" N; h = 2362.9 m

Determine if an obstruction truly exists. If the top of Maiden Peak doesn't cause an obstruction, do you think additional observations should be taken on the mountain to determine if an obstruction exists? Discuss.

5. Your job is to set a point (foresight point) having geodetic coordinates shown below. Using a total station and the two points (occupied point and backsight point) given, determine the horizontal angle and horizontal distance which you would use to set the point. Also, how would you determine where to set the point vertically?

POINT	λ	Φ	h
Occupied	121°47'06.11898"W	42°15'22.59645"N	1302.365 m
Backsight	121°47'34.32574"W	42°15'15.61009"N	1289.871 m
Foresight	121°47'12"W	42°15'20"N	1300.000 m

6. A baseline is measured with an EDMI at 2,467.782 meters. The average elevation of the line is 1220 meters on the NAVD88 datum. The average geoid height for the line is -25.910 meters. What is the length of the line on the ellipsoid surface? What is the computed change in length of line due to in this case?

CHAPTER ELEVEN
Satellite Coordinate Systems

An understanding of the relationship between orbiting satellites and terrestrial stations requires the use of both space-fixed coordinate systems and earth-fixed coordinate systems. We have already described and used two earth-fixed coordinate systems: geodetic coordinates and geocentric coordinates. This chapter explores a space-fixed or inertial frame to be used for describing satellite positions with respect to time. We must investigate some aspects of geodetic astronomy in order to define such a space-fixed reference frame.

Coordinate Confusion

The earth-fixed coordinate systems we have used are extremely handy for describing the positions of terrestrial stations. With the exception of tectonic motion, the positions are fixed with respect to time. It would be quite confusing if the coordinates of earth stations changed with the rotation of the earth about its spin axis! Now consider a satellite orbiting the earth as shown in figure 11.1. Not only does the satellite's position in its orbital plane change with respect to time, but also its position with respect to an earth-fixed coordinate system is constantly changing. This latter characteristic would be true even if the satellite somehow ceased orbiting and was "suspended" in space.

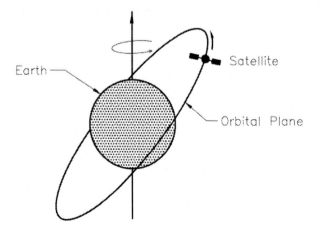

Figure 11.1 Satellite orbiting earth.

The determination of terrestrial positions using global positioning system (GPS) satellites is facilitated by the use of an earth-centered, space-fixed coordinate system. A stationary observer in this reference frame will see both the terrestrial station and the satellite moving with respect to time; however, each motion will be rather simple to describe mathematically. This space-fixed coordinate system will be used to transform satellite positions into terrestrial positions.

The Celestial Sphere

Consider a sphere having its origin at the earth's center of mass and having an infinite radius. This *celestial sphere* is fundamental to geodetic astronomy. Now, extend the earth's actual rotation axis to intersect the celestial sphere. The *celestial ephemeris pole* (CEP) is located at the point where the north end of the rotation axis intersects the celestial sphere. The CEP lies at the positive end of the **Z**-axis for the *celestial coordinate system*. To describe the position of the **X**-axis for this system, it is necessary to understand something of the earth's orbit about the sun.

The Earth's Orbit

The earth orbits about the sun in a planar elliptical path. This orbital plane is known as the ecliptic. The sun is located at one of the foci of the elliptical orbit. The earth makes a complete orbit of the sun during a sidereal year (sidereal time is described later in this chapter). That point of the orbit lying nearest to the sun is known as the perihelion and that point lying farthest from the sun is the aphelion. Presently, the earth passes through the perihelion on January 3 and the aphelion on July 3 (Vanicek and Krakiwsky, 1982). These dates nearly correspond with the winter and summer solstices, respectively. Figure 11.2 is a depiction of the ecliptic.

The earth's spin axis, denoted as the CEP in the figure, is not oriented perpendicular to the ecliptic, rather it is skewed at an angle of approximately 23.5° from the perpendicular. This angle is known as the *obliquity* (ϵ). Figure 11.3 shows the earth's obliquity. Were it not for the obliquity, the days and nights would be the same length (12 hr.) throughout the year. The obliquity also defines the most northerly and southerly latitude where the sun will be

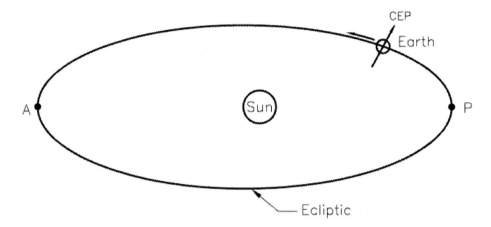

Figure 11.2 The ecliptic.

directly overhead at noon (review the work of Eratosthenes in Chapter 1). These latitudes are denoted as the Tropics of Cancer and Capricorn on the globe. The perpendicular to the ecliptic, containing the earth's center of mass, is known as the *north ecliptic pole* (NEP).

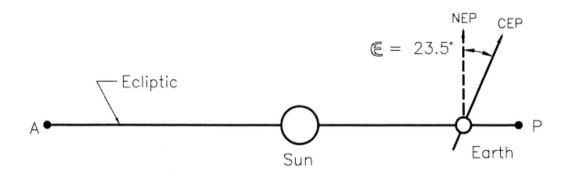

Figure 11.3 The Earth's Obliquity.

The International Celestial Reference Frame

The **X**-axis for the space-fixed International Celestial Reference Frame (ICRF) is defined by the intersection of the celestial equatorial plane with the ecliptic. The positive end of this axis is directed toward the sun at the vernal equinox in the northern hemisphere. The *First Point of Aries* (denoted ϒ) is located along the positive **X**-axis at its intersection with the celestial sphere.

The **Y**-axis forms a right-handed orthogonal coordinate system with the **X**-axis and the CEP (positive **Z**-axis). Figure 11.4 illustrates the ICRF and its relationship with the NEP.

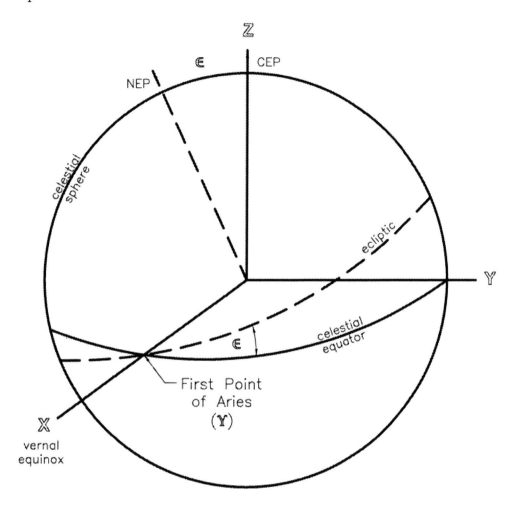

Figure 11.4 Space-fixed International Celestial Reference Frame.

The ICRF was determined using radio positions of 212 extragalactic sources determined through very long baseline interferometry (VLBI). The positional accuracy of the sources is said to be better than 1 milli-arc second (mas). The ICRF was defined by the International Astronomical Union (IAU) and adopted as of January 1, 1998.

Star positions are catalogued with respect to the ICRF using measurements analogous to earth-fixed spherical latitude and longitude. Star positions are expressed using right ascension (α) and declination (δ). Right ascension is measured in the equatorial plane from the positive X-axis toward

197

the positive Y-axis. Declination is measured from the celestial equatorial plane. This is shown in figure 11.5.

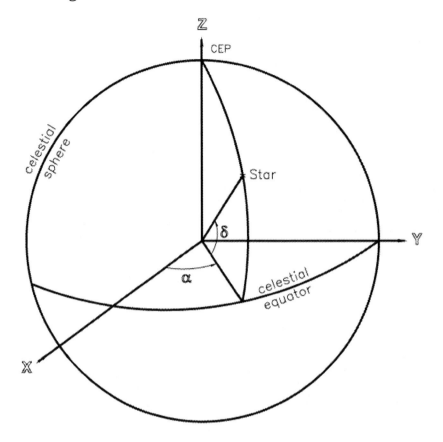

Figure 11.5 Cataloguing star positions.

Motion of the Earth's Axis

The orientation of the earth's rotation axis changes over time with respect to the stars. The celestial motions that cause these changes may be explained by the physics of this dynamic system. These motions are precession and nutation.

Precession

The earth's obliquity causes a motion known as *solar precession* that is likened to the motion of a gyroscope's rotation axis while the gyroscope is spinning. The rotation axis also precesses in response to the gravitational attraction from the moon and other planets, although to a much lesser extent

198

than solar precession. The earth's spin axis forms a circular cone (figure 11.6) as a result of precession, its orientation with respect to the stars constantly changing. The rate of precession is approximately 50 arc-seconds annually (Mackie 1985). Accordingly, the period of the precession is approximately 26,000 years.

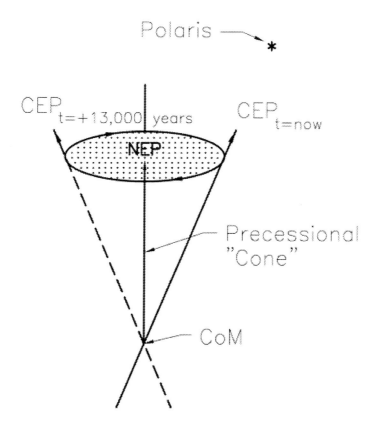

Figure 11.6 The earth's precession.

The cause of solar precession can be discovered by examining figure 11.7. We may consider the earth as having two "hemispheres" with respect to the ecliptic, a "nearer hemisphere" lying closer to the sun and a "farther hemisphere" lying away from the sun. Newton's law of gravitation tells us there will be a gravitational attraction between each "hemisphere" and the sun with the "nearer hemisphere" feeling a greater attraction than the "farther hemisphere". The inequality of these two forces causes a torque about an imaginary axis that is perpendicular to the sun and running through the earth's center of mass. This torque seeks to "right" the earth's orientation and is responsible for precession.

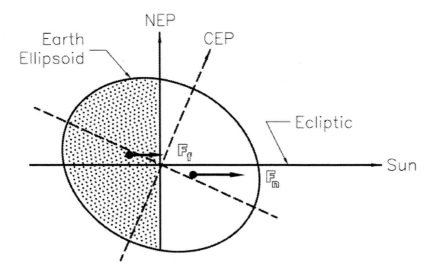

Figure 11.7 Cause of solar precession.

Nutation

The distance between the earth, sun, moon and planets is constantly changing with respect to time due to various orbits. The gravitational attraction between the earth and these other celestial bodies changes with the changing distances. These variations in gravitation cause a perturbation of the earth's

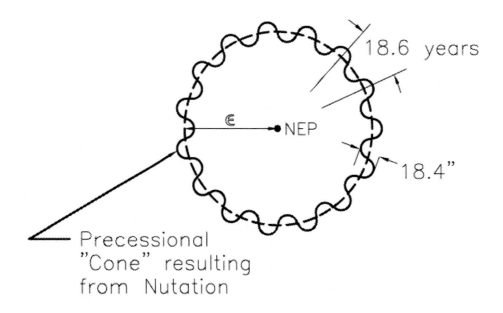

Figure 11.8 Nutation.

precession known as nutation. Nutation causes the precession of the rotation axis to be irregular, instead of smooth. Figure 11.8 illustrates the regular effect of nutation on the precession of the earth's axis of rotation. The period of the nutation is 18.6 years.

Celestial Pole Offsets

The International Earth Rotation Service (IERS) tracks and publishes Earth Orientation Parameters (EOP). The effects of precession and nutation are quantified using celestial pole offsets. The celestial pole offsets describe the position of the celestial pole with respect to its position defined by the conventional International Astronomical Union (IAU) precession/nutation models.

Polar Motion

Geodetic astronomers observing stars at a fixed latitude detected an unusual phenomenon. Using a photogrammetric zenith tube (PZT) to establish the zenith at a particular location, the astronomers were able to observe stars at or near the PZT zenith. Observations taken during the course of a night would result in a map of the stars at the PZT's latitude. Over a period of time, it was observed that the positions of the stars varied slightly; i.e., a star located previously at the exact PZT zenith would no longer be at the exact zenith. The star appeared to have shifted.

These observations, along with others, provided evidence of *polar motion.* Polar motion is movement of the rotation axis of the earth with respect to the crust. Not well-understood, polar motion is a well-established phenomenon that has an amplitude of less than 10 meters. The period of polar motion is about 434 days (Leick 1995). The International Earth Rotation Service plots the location of the CEP with respect to time. Such a plot is shown in figure 11.9.

Polar motion is described as a "wobble" in the rotation axis in response to *free nutation* that accompanies any gyroscopic motion (Vanicek and Krakiwsky, 1982). The viscous nature of the earth's interior further complicates this motion. There are three primary components of polar motion.

The first, known as the *Chandler wobble*, is a free oscillation having a period of about 435 days. A second annual oscillation is forced by the seasonal

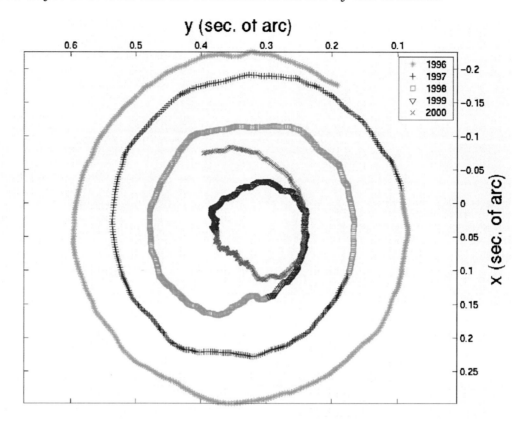

Figure 11.9 Plot of polar motion (courtesy U.S. Naval Observatory).

displacement of air and water masses. The third component is found in the irregular drift of the mean pole. Polar motion cannot be completely described either mathematically or physically though it has been the subject of much research during the past 100 years.

Terrestrial Poles

The desirability of having a "fixed" position for the North Pole led to the establishment of the *conventional international origin* (CIO) that was defined as the mean position of the North Pole as determined by observations taken between 1900 and 1905. This fixed, historic pole position, became known as the *conventional terrestrial pole* (CTP) and was the basis for the rotational axis (z-axis) of the previously described earth-fixed geocentric coordinate system known as the *conventional terrestrial reference system* (CTRS). It is critical that a terrestrial reference system have a fixed pole position so that positions of terrestrial stations remain fixed with respect to time.

The International Earth Rotation Service (IERS) adopted the International Terrestrial Reference System (ITRS) in 1996. The IERS Reference Pole (IRP) and the IERS Reference Meridian (IRM) define the orientation of the ITRS. The IRP corresponds with the direction of the CTP with an uncertainty of 0.005". The IRM is not exactly coincident with the Greenwich meridian. The IERS defines coordinate systems or *reference frames* based upon the ITRS with respect to time. The ITRS has several realizations (See Chapter 12 Geodesy and GPS).

The polar motion graph in figure 11.9 has as its origin the IRP. The graph is simply the location of the CEP with respect to the IRP as a function of time. The CEP position is described using (x_p, y_p) coordinates where the x_p axis is coincident with the plane of the IRM and the positive y_p axis is coincident with the $\lambda = 270°$ meridian plane. Note that the CEP represents the actual, time-dependent location of the earth's spin axis.

Rate of Rotation

The earth's rate of rotation, while nominally accepted as one rotation per day ($2\pi/86,400$ s) varies in response to secular, periodic and irregular effects (Vanicek and Krakiwsky, 1982). The secular (linear) effect is a continuous slowing of the earth's spin rate of approximately 0.002 s per century. This lengthening of a day is theorized to be the result of tidal friction. Periodic variations are thought to be the result of tidal and wind effects (Vanicek and Krakiwsky, 1982) while other periodic effects are thought to be related to expansion and contraction of the atmosphere. These periodic variations may be as great as several milliseconds over their periods that vary from one year to a month. Unexplained, irregular effects on the spin rate may account for changes of 10 milliseconds per day.

Time

Time has its basis in the earth's rotation with respect to the sun and the stars. Two different general time systems have developed as a result of these two references: solar time and sidereal time.

Solar Time

Solar time is measured with respect to the sun. *Apparent solar time* is based upon the earth's actual, varying spin rate. *Mean solar time*, or *civil time*, is based upon a mean solar day, i.e., one earth revolution per 86,400 seconds. The difference between apparent and mean solar time is known as the *equation of time*. The value of the equation of time ranges between ±16 minutes of time.

Sidereal Time

Sidereal time or "star time" is measured with respect to the stars. The relationship between solar time and sidereal time can be explained using figure 11.10. The earth orbits the sun in 366.2564 sidereal days (Vanicek and Krakiwsky, 1982). A sidereal day being the time required for a chosen star to cross the observer's meridian successively, i.e. an earth rotation with respect to the stars. The earth orbits the sun in 365.2564 solar days (Vanicek and Krakiwsky, 1982). Why the difference?

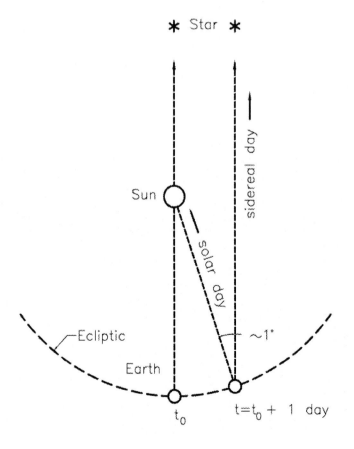

Figure 11.10 Difference between a Solar and a Sidereal Day.

Because the earth is orbiting the sun as it revolves, the earth must spin farther, with respect to the stars, in order for the sun to cross the observer's meridian after one day. At epoch t_0 in figure 11.10, the observer's meridian crosses both the sun and a selected star. After one sidereal day, the observer's meridian crosses the same star, but due to the earth's orbit about the sun it does not cross the sun. The earth must rotate an additional amount (approximately 4 minutes of time or 1 degree of arc) in order for the sun to once again cross the observer's meridian. Thus, a sidereal day is shorter than a solar day.

Greenwich Apparent Sidereal Time

The earth-fixed ITRS rotates with respect to the space-fixed celestial coordinate system. The position of the x-axis (IERS Reference Meridian) for the ITRS is described with respect to the **X**-axis of the ICRF using Greenwich Apparent Sidereal Time (GAST). GAST represents an angle, analogous to a "time-dependent longitude", measured in the celestial equatorial plane. Figure 11.11 depicts GAST and the time-dependent polar motion coordinates x_p and y_p.

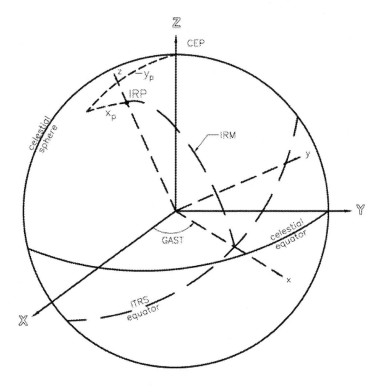

Figure 11.11 Relationship between ITRS and ICRF.

Earth Orientation

The position of a satellite may be readily expressed in the ICRF. In order to use satellites for the determination of terrestrial positions, it is necessary to transform coordinates in the ICRF to ITRS coordinates. This *earth orientation* transformation is handled using three-dimensional coordinate transformations as described in Chapter 7. The transformation may be expressed simplistically as $\mathbf{X}_{\text{CTRS}} = \mathbf{R}_{\text{PM}}\mathbf{R}_{\text{GAST}}\mathbf{R}_{\text{N}}\mathbf{R}_{\text{P}}\mathbf{X}_{\text{ICRF}}$

where $\mathbf{X}_{\text{ITRS}} = \begin{bmatrix} x \\ y \\ z \end{bmatrix}$ = earth-fixed, geocentric coordinates.

$\mathbf{R}_{\text{PM}}$ = rotation matrix for polar motion.

$\mathbf{R}_{\text{GAST}}$ = rotation matrix for Greenwich apparent sidereal time.

$\mathbf{R}_{\text{N}}$ = rotation matrix for nutation.

$\mathbf{R}_{\text{P}}$ = rotation matrix for precession.

$\mathbf{X}_{\text{ICRF}} = \begin{bmatrix} \mathbf{X} \\ \mathbf{Y} \\ \mathbf{Z} \end{bmatrix}$ = space-fixed ICRF coordinates.

Equation of Motion

The equation of motion for an orbiting satellite can be readily derived from Newton's laws of inertia and universal gravitation presented in Chapter 4. The equation describing the normal (unperturbed) motion of an orbiting satellite is $\ddot{\mathbf{X}} = \dfrac{-GM}{|\mathbf{X}|^3}\mathbf{X}$ where $\ddot{\mathbf{X}}$ is the acceleration vector of the satellite, G is the universal gravitation constant, M is the mass of the earth, and $\mathbf{X}$ is the position vector of the satellite with respect to the earth's center of mass. The equation assumes that the earth's mass is located at a point, the center of mass.

The State Vector

The *state vector* describes the position and velocity of a satellite at an epoch in time. The ICRF is the reference frame for the satellite state vector.

Figure 11.12 depicts the position vector (**X**) and velocity vector ($\dot{\mathbf{X}}$) for a satellite. Note that the velocity vector is always perpendicular to the position vector. The position vector may be written $\mathbf{X} = [\mathbf{X}\ \ \mathbf{Y}\ \ \mathbf{Z}]^{\mathrm{T}}$ and the velocity vector

$$\dot{\mathbf{X}} = \begin{bmatrix}\dot{\mathbf{X}} & \dot{\mathbf{Y}} & \dot{\mathbf{Z}}\end{bmatrix}^{\mathrm{T}}$$

The state vector represents six variables that are a function of time. Accordingly, the state vector changes at each epoch in time. Since a GPS satellite, for instance, moves at a speed of approximately 4 km/sec, this time dependency may be considered an undesirable characteristic of the state vector. A method of describing the state of a satellite using fewer time dependent variables is desirable. Keplerian orbital elements provide a simpler way!

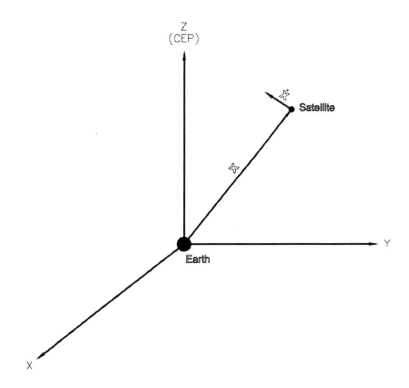

Figure 11.12 Satellite state vector.

Kepler's Laws

The orbital characteristics of satellites were ascertained and described by the German astronomer Johannes Kepler approximately 400 years ago. The three laws of planetary motion postulated by Kepler also govern orbiting satellites – whether they be communications satellites or GPS satellites. Kepler, a mathematician and astronomer, based his laws on planetary data collected by Danish astronomer Tycho Brahe. Kepler's laws are exact for unperturbed orbits. Sir Issac Newton relied heavily on Kepler's work in forming his law of gravitation.

Kepler's First Law

Kepler's first law states that *the orbit of each planet about the sun is an ellipse with the sun located at one focus.* Similarly, the normal (unperturbed) orbit of a satellite about the earth is elliptical, with the earth located at a focus. Figure 11.13 shows the orbital ellipse for a satellite. *Perigee* ("near earth") is the location where the satellite is nearest the earth while *apogee* ("far earth") is the satellite's farthest location from the earth. Perihelion and Aphelion are the corresponding terms for the planetary orbits about the sun.

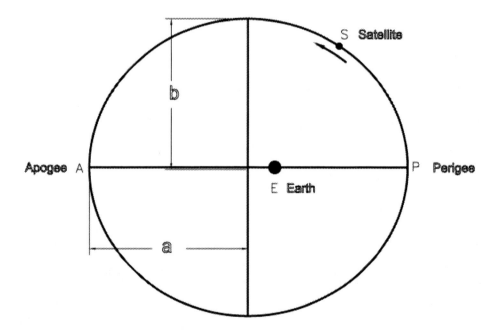

Figure 11.13 Satellite orbital plane.

Kepler's Second Law

The line from the center of the sun to any planet sweeps out equal areas of space in equal time is Kepler's second law. The ramification of this law is that planets in lower level orbits (nearer the sun) must orbit more rapidly than planets in higher level orbits. Kepler's second law is illustrated using figure 11.14. Again, this law holds for normal satellite orbits as well as planets. According to this law, if $(t_2 - t_1) = (t_4 - t_3)$ then $A_1 = A_2$ where t represents an epoch and A represents area. This law may also be stated as a rate of area change using

$$\frac{dA}{dt} = \text{constant} = \frac{1}{2}r^2 n$$ where r is the range from the earth's center of mass to the satellite and n is the *mean motion* (or angular velocity) of the orbiting satellite. The mean motion is typically expressed in radians per second using the orbital period, T, $n = 2\pi/T$. The differential area, dA, could also be expressed in terms of the *true anomaly*, v, using dA = ½ r^2dv.

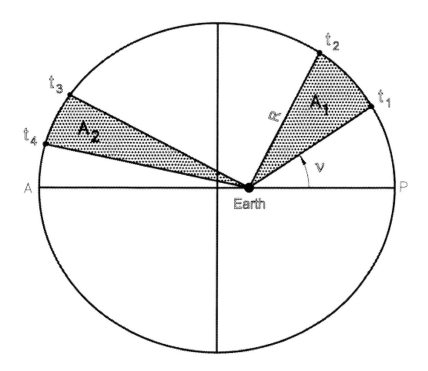

Figure 11.14 Kepler's second law.

Kepler's Third Law

Kepler's third law states *the cubes of the semi-major axes of the planetary orbits are proportional to the squares of the planets' periods of revolution.* This law may be stated using the equation a^3/T^2 = constant where a is the semi-major axis of the planetary (or satellite) orbit and T is the orbital period. Should one substitute for the orbital period using the mean motion, we find that $a^3n^2/4\pi^2$ = constant so that a^3n^2 = constant = GM. The proof for this is left to others.

Normal Satellite Orbits

Normal (unperturbed) satellite orbits exhibit the following properties:

1. Follow Kepler's laws.
2. Follow a planar elliptical orbit.
3. The orbital plane contains the earth's center of mass.
4. The orbital plane is fixed with respect to the stars (and the ICRF).

Keplerian Orbital Elements

Six parameters or *Keplerian orbital elements* are used to describe the state of a satellite in its orbit. The orbital elements for a satellite determine when and where a satellite will be visible to users. This is of critical importance

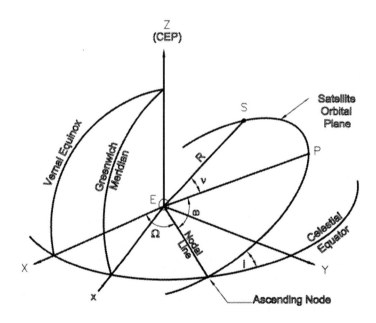

Figure 11.15 Keplerian orbital elements.

for users of the global positioning system (GPS). Describing the state of a satellite using Keplerian orbital elements has the advantage over the state vector in that only one of these six orbital elements is a function of time. Figures 11.15 and 11.16 are used to illustrate the orbital elements. The elements are as follows:

1. Semi-major axis of the orbital ellipse, a.

2. First eccentricity of the orbital ellipse, e = [(a² − b²)/a²]^{1/2}.

3. *Inclination of the orbital plane*, i. The inclination angle is measured relative to the celestial equatorial plane as shown in figure 11.15.

4. *Right ascension of the ascending node*, Ω. The ascending node is that point where the satellite intersects the celestial equatorial plane on its northerly track. The intersection of the celestial equatorial plane with the orbital plane is known as the *nodal line*.

5. *Argument of perigee*, ω, is that angle, measured in the orbital plane, from the nodal line to the major axis of the orbital ellipse in the direction of the satellite motion.

6. *True anomaly*, ν, is that angle, measured in the orbital plane, from the major axis to the satellite in the direction of the satellite motion.

Of these orbital elements, only the true anomaly is time-dependent.

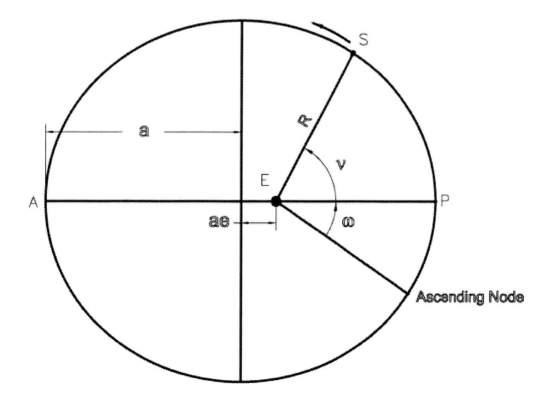

Figure 11.16 Satellite orbital plane showing Keplerian orbital elements.

Keplerian Coordinate System

The orbital plane of a satellite may also be used to establish a rectangular coordinate system that is needed for the transformation from a satellite position expressed in Keplerian orbital elements to the ICRF. This rectangular system, referred to as the *Keplerian coordinate system* (x_k, y_k, z_k) is shown in figure 11.17.

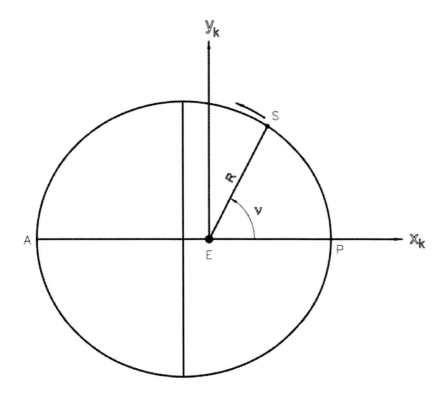

Figure 11.17 Keplerian coordinate system

Coordinates for an orbiting satellite may be computed using the following functions:

$$x_k = r \cos v$$
$$y_k = r \sin v$$
$$z_k = 0$$

Transformation to the celestial coordinate system is accomplished using

$$\begin{bmatrix} X \\ Y \\ Z \end{bmatrix}_{ICRF} = \mathbf{R}_3(-\Omega)\,\mathbf{R}_1(-i)\,\mathbf{R}_3(-\omega) \begin{bmatrix} x_k \\ y_k \\ z_k \end{bmatrix}$$

212

Orbital Perturbations

Our assumption of a normal satellite orbit, like Kepler's assumption of planetary orbits, is somewhat simplistic. There are a number of accelerations acting upon orbiting satellites that cause their orbits to be disturbed or *perturbed*. These accelerations are all dependent upon the position of the satellite within its orbit. Some of these orbital perturbing accelerations are listed below.

1. Accelerations due to a non-spherical earth and inhomogeneous mass distribution within the earth.
2. Accelerations due to other celestial bodies, primarily the sun and moon.
3. Accelerations due to earth and oceanic tides.
4. Acceleration due to atmospheric drag, primarily affecting low-earth orbit satellites.
5. Accelerations due to direct and earth-reflected solar radiation pressure.

These accelerations are added together to yield a vector of perturbing accelerations. This vector must then be added to the acceleration describing the motion of a satellite in order to get an accurate picture of the perturbed satellite orbit. These perturbations also cause a velocity (change with respect to time) in orbital **elements** Ω and ω. Thus, the orbital plane of a perturbed satellite is not fixed within the ICRF.

Study Questions

1. Assume that polar motion coordinates at a particular epoch are x_p = 0.256" and y_p = -0.182". Compute the GRS 80 geodetic distance from the IRP to the CEP.
2. Determine the time difference, to the nearest second, between a mean solar day and a mean sidereal day.
3. Determine the rotation axes and magnitude and sense of the rotations for $\mathbf{R}_{PM}$ and $\mathbf{R}_{GAST}$.
4. Determine the mean motion of a GPS satellite (in rad/sec) if the orbital period is 12 sidereal hours. Determine the semi-major axis of the orbital ellipse using Kepler's third law. Determine the approximate speed (in km/sec) of the satellite.

5. Determine the new period for a satellite in terms of its old period, T_{old}, if the orbit is increased by a factor of two.

6. Plot the ground track (latitude v. longitude) of a GPS satellite having a period of 12 sidereal hours and an inclination of 55°. Assume that the satellite begins at $\varphi = 0°$, $\lambda = 0°$ and makes three orbits around the earth. (HINT: Consider the relative motion of the earth and the satellite, viewed from the North Pole.) Show your calculations of the satellite position with respect to time.

7. Repeat the previous study question for a "retrograde GPS satellite" having an inclination of 125° and a period of 12 sidereal hours.

CHAPTER TWELVE

GEODESY AND GPS

The most important development in the history of surveying occurred in the 1980's when NAVSTAR Global Positioning System (GPS) became operational. Surprisingly, few in the profession were aware it was about to happen.
Dracup, Geodetic Surveying 1940-1990

The Global Positioning System changed everything. The beginning of the new millennium is the story of the evolution of three modern three-dimensional (3-D) reference systems: NAD 83, WGS 84, and the International Terrestrial Reference System (ITRS). Snay and Soler (1999) review the steps to the establishment of a modern 3-D reference system.

1. The origin of the modern 3-D reference system is at the center of mass of the earth. The z-axis should include the International Earth Rotation Service (IERS) Reference Pole (IRP). The IERS is an international organization established in 1988 and headquartered in Paris, France. The x-axis should be placed through the point of zero longitude located on the plane of the conventional equator as defined by IERS. The y-axis should form a right-handed coordinate system with the x and z axes. These axes are linked to physical locations (monuments) on or within the surface of the earth.

2. The unit of length gives us the concept of distance and coordinates on these coordinate axes. The meter is defined as the distance traveled by electromagnetic energy in a vacuum during the time interval of 1/ 299,792,458 seconds (exact). Distances measured by GPS and technologies such as electro-optical distance measuring instruments (EDMI), Doppler satellite observations, very-long baseline interferometry (VLBI) and satellite laser ranging (SLR) are difficult to calibrate exactly. Therefore, the measurements contain uncertainties. This is a problem because scale adjustments cause coordinate changes. Accordingly, reference frames may differ slightly in scale, particularly where terrestrial measurements (EDMI) were incorporated into the datum realization.

3. The ellipsoid of revolution or reference surface used by the modern 3D coordinate system must be defined. The geometric center of the ellipsoid is located at the origin of the 3-D coordinate system. The semi-minor axis of the ellipsoid must coincide with the Cartesian z-axis.

These first three steps provide the datum definition. According to Strange (2000), "A datum is defined by specifying in words the location and orientation relative to the Earth of the datum axes, the scale to be used, and size and shape of the reference ellipsoid." Though not necessary to define a geometric datum, a fourth step may be used to tie the geometric reference system to the physical reference system.

4. The use of the gravity field of the earth to define the relationship between ellipsoid height and orthometric height.

The task of defining a reference system is exacerbated by the earth's dynamic behavior as described in Chapter 11. The earth's center of mass is moving with respect to the earth's surface. Variations can be measured in the earth's rotation rate. The pole position changes by polar motion. The earth's rotation axis is not fixed, but varies due to precession and nutation. In fact, points on the earth's surface are all moving with respect to each other due to:

- Plate tectonics - The earth's crust is made up of some 20 tectonic plates, rigid in nature, all moving with respect to each other. This movement can approach 150 millimeters per year, and is easily measurable using GPS technology.
- Earthquakes
- Volcanic activity
- Post glacial rebound
- Subsidence caused by the extraction of underground fluids
- Solid earth tides

The effect of the earth's dynamic behavior can be captured by introducing a temporal dimension into the reference system. The International Earth

Rotation Service (IERS) introduced the International Terrestrial Reference Frame of 1988 (ITRF88) to support scientific activities such as monitoring crustal motion and polar motion. Positions and velocities are the published coordinates for monuments in this reference system.

The maintenance of the NAD 83, WGS 84, and ITRS reference systems is an evolving process requiring new sets of published coordinates over time. Each set of published coordinates is the usable product of a particular "realization" or particular reference frame of the reference system such as NAD 83. Strange (2000) describes the process of datum realization: "Realizing a datum consists of computing and making available to users, coordinates of two types of reference points, time varying satellite locations (orbits), and monumented reference stations.

To realize a datum, equations are developed and used to compute reference point coordinates from observations. Geodetic organizations try to develop equations that will provide reference point coordinates relative to the datum as defined. But these equations are never perfect and there can be systematic error in the computed coordinates of reference points relative to the datum as defined. This systematic error is best viewed as meaning that the datum, as realized by reference point coordinates, differs from the datum as defined. With current datums, the defined and realized datums differ only by the order of a centimeter and the distinction is not usually important. But at the time WGS 84 and NAD 83 were first realized, this difference was about two meters (though they were thought to be identical at the time). The datum as realized is the relevant datum for users. It is the only datum to which they have access." This chapter examines different realizations of NAD 83, WGS 84, and the ITRS.

NAD 83 (1986) – The 1st Realization

The initial realization of the North American Datum of 1983 was completed in 1986 without the benefit of the Global Positioning System. Traditional methods of triangulation, trilateration, and traverse provided the bulk of the observations for the gigantic project that was NAD 83. The exception was the use of Doppler satellite observations from the U.S. Navy's TRANSIT system to determine the orientation/scale of the 3D Cartesian coordinate axes and the global fit of the GRS 80 ellipsoid. Doppler-derived positions were transformed so that they would be consistent with measurements made using VLBI, SLR and terrestrial azimuths.

While theoretically a 3-D datum, the lack of space-derived heights on control monuments rendered NAD 83 (1986) accurate for horizontal (latitude and longitude) positions only. Still, NAD 83 (1986) was a bold departure from NAD 27, the horizontal datum it replaced. Satellite observations from the TRANSIT System enabled a global viewpoint of the earth. Doppler shift is a physical phenomenon where the frequency of an acoustic signal varies as a function of velocity between the source of the signal and an observer. It was determined that measurement of the Doppler shift used in conjunction with the known positions of the TRANSIT satellites could provide the positions of receivers on the earth's surface. The determination of a few hundred positions on the earth's surface benefited the development of NAD 83. The TRANSIT system was made obsolete by the impact of the NAVSTAR GPS system in the 1990's.

Unfortunately, the revolutionary impact of the NAVSTAR GPS system was unforeseen during the development of NAD 83. It must have been very difficult for those who had worked so hard on the implementation of NAD 83 to have its usefulness questioned for project control work. The emerging GPS survey user community experienced two significant problems with NAD 83 (1986):

1. The use of GPS in the relative positioning mode routinely provided precise baseline vectors with a relative precision of 1: 1,000,000. The positional tolerance of NAD 83 control monuments was approximately 1: 100,000. Surveyors desired control monuments with a relative precision as good as or better than their observed GPS baselines.

2. The existing placement of NAD 83 control monuments favored the original need of triangulation methods. Triangulation involved the measurement of the angles of large geometric figures (triangulation arcs) on the earth's surface (see figure 12.1). To observe the angles between

control points located many miles away, it was necessary to use unobstructed terrain points such as mountain tops or to build steel observation towers (Bilby towers) 100 - 150 feet high where the land was flat. Dracup states: "As for Bilby towers, the last erected by the National

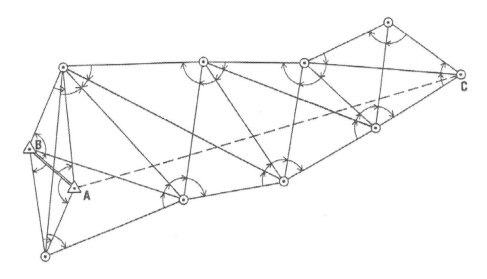

KNOWN DATA:
    Length of base line AB.
    Latitude and longitude of points A and B.
    Azimuth of line AB.

MEASURED DATA:
    Angles to new control points.

COMPUTED DATA:
    Latitude and longitude of point C, and other new points.
    Length and azimuth of line AC.
    Length and azimuth of all other lines.

Figure 12.1 Typical Triangulation network. Geodesy for the Layman, National Ocean Service, July 1985, pg. 15.

Geodetic Survey was in September 1984 at a station appropriately named BILBY, near Hartford, CT." Since the use of GPS was unforeseen, conventionally observed monuments are often difficult to access or possess obstacles to the receipt of satellite signals.

The mid-1980's marked the end of triangulation and trilateration as the method of providing geodetic control products. GPS technology became the tool for horizontal positioning. Individual states, in collaboration with the National Geodetic Survey, used GPS technology to establish statewide control networks termed "High Accuracy Reference Networks" (HARNS). The state of Tennessee possessed the first HARN in 1989. Many of the HARN control points were existent NAD 83 control monuments. Therefore, it became possible to have more than one set of published coordinates for the same monument. This illustrates the concept of more than one realization of the same datum.

NAD 83 (HARN) – The 2nd Realization

The use of GPS and associated technologies such as satellite laser ranging (SLR), and very-long baseline interferometry (VLBI) indicated that the origin of the NAD 83 (1986) reference system had not been fitted to the exact center of mass of the earth. It was off by approximately 2 meters. Also the orientation of the 3-D coordinate axes was misaligned by approximately 0.03 arc-seconds, and that the scale of the axes differed some -0.0871 ppm from the true definition of the meter.

During the adjustment of each state HARN, the National Geodetic Survey elected to retain the original geocenter of NAD 83 (1986) and the alignment of the 3-D coordinate axes, but did adjust the scale. This scale adjustment (-0.0871 ppm) had an impact on absolute horizontal positions (latitude and longitude) of approximately a meter. However the scale adjustment systematically decreased ellipsoid heights approximately 0.6 meters computed as (-0.0871E-6 * 6,371,000 meters). This adjustment for scale has facilitated research and development toward using GPS technology for determining precise ellipsoid heights and orthometric heights.

The important point in considering the 2nd realization of NAD 83 is that it did not happen overnight. Indiana was the last state to act in creating a state-wide HARN in the late 1990's. The era of the state HARN's spanned from 1989 to 1998. The GPS equipment available and procedures used varied during this long period. Research and development toward achieving reliable ellipsoid heights was also accomplished during this period. A penalty for the piecemeal growth of the HARN's may be their considerable variability in accuracy with respect to the evolving NAD 83 reference frame.

The state HARN's, or as Snay and Soler (February 2000) term "the 2nd realization of NAD 83", did meet the GPS survey user community demand for precise control monuments between which to fit GPS baselines with minimal distortion. Some HARN stations are newly monumented points while others were existent NAD 83 terrestrial stations. These latter stations now possess coordinates expressed in two realizations of NAD 83 – the original (1986) adjustment and the HARN adjustment. The HARN adjustment positions being updated, more precise, coordinates in the NAD 83 evolving reference frame. Both of these positions are maintained in the NSRS database. The most recent position is typically labeled "Current Survey Control" on NGS data sheets while older positions are labeled "Superseded Survey Control".

NAD 83 (CORS93) – The 3rd Realization

A CORS station is a fixed reference station comprised of a GPS receiver, antenna assembly, personal computer, and communications link for real-time applications. The motivation for the early growth of the Continuously Operating Reference Stations (CORS) was to provide real-time differential positioning in support of marine and air navigation. The U.S Coast Guard, U.S Army Corps of Engineers, and the Federal Aviation Administration invested in CORS stations to provide navigation services. As a result, many stations are sited at airports and along waterways and coastlines. The current national CORS network is shown in figure 12.2.

The principle behind differential positioning is that nearly all GPS receivers develop similar errors tracking the same signals from the local satellite constellation through similar columns of the atmosphere. Place one of the receivers on a known position. This fixed receiver is then used to calculate the magnitude of error and broadcast this error computation as a correction to the other receiver(s). CORS stations can provide these corrections 24 hours a day 365 days a year in support of navigation and real-time positioning. While the primary mission of many CORS stations is navigation, the federal government has adopted the policy that these stations can provide other services, such as geodetic positioning.

Another motivation for the development of CORS stations is that each station represents one less receiver a government agency or contractor must procure. The typical CORS station collects data 24 hours a day at predefined time intervals (epochs). Periodically this data is converted from the receiver manufacturer's proprietary format into RINEX (receiver-independent exchange) format. This data is available on the Internet for post processing

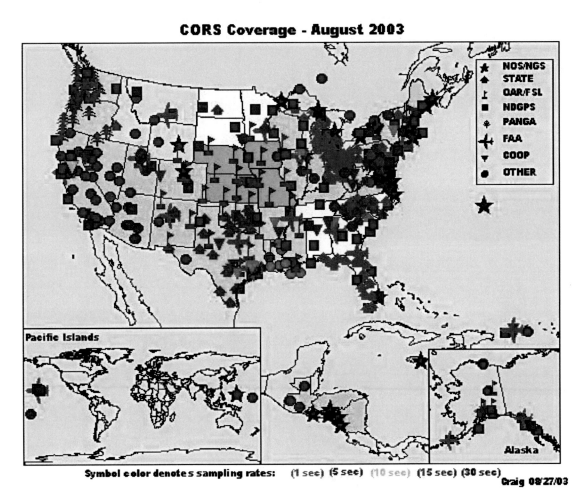

Figure 12.2 U.S. National CORS network. www.ngs.noaa.gov/CORS.

(www.ngs.noaa.gov/CORS). Assuming a construction site is surrounded by three operating CORS stations within a reasonable distance, it would be possible for a one-person survey crew to occupy the unknown project control points. Since all receivers were operating simultaneously, the Internet data could be used to compute positions for the points.

The geodetic implication of continuously operating receivers is that the element of time is introduced. The crustal motion of the earth will change the position of the operating CORS station over time. This encourages the computation of CORS station positions in the International Terrestrial Reference System (ITRS). Since the International Earth Rotation Service publishes revised positions for its stations almost every year, annual realizations are not unexpected.

The positions of early CORS stations were computed in the ITRF93 reference frame. The parameters for a coordinate transformation from ITRF93 to NAD 83 (CORS 93) were computed from the known positions of nine VLBI stations in the United States. This resulted in an adjustment in scale of approximately 0.005 ppm translating to a discrepancy in position between NAD 83 (HARN) and NAD 83 (CORS93) of approximately ten centimeters in the horizontal although a rigorous transformation between NAD 83 (HARN) and later NAD 83 realizations is not readily available.

NAD 83 (CORS94) – The 4th Realization

CORS station positions were next computed in the ITRF94 reference frame. The parameters for the coordinate transformation to NAD 83 (CORS94) were computed from eight of the nine U.S. VLBI stations. The ninth station located in California was dropped due to concerns with crustal motion. While the majority of the continental United States is located on one tectonic plate, the North American Plate, California west of the San Andreas Fault is located on another tectonic plate, the Pacific Plate. The NAD 83 (CORS94) reference frame can technically be termed the fourth realization of NAD 83.

NAD 83 (CORS96) – The 5th realization

Positions of CORS stations that became operational after the fall of 1998 were computed in ITRS96. The parameters of the coordinate transformation from ITRS96 to NAD 83 (CORS96) are computed from the previously referenced eight U.S VLBI stations plus four Canadian VLBI stations. Positional discrepancies between NAD 83 (CORSxx) reference frames are less than 2 centimeters horizontally and 4 centimeters vertically. Recent re-observations of HARN's provide station positions in this realization of NAD 83.

The realizations of NAD 83 hold the original fit of the origin to the center of mass of the earth and the original misalignment of the 3-D coordinate axes. The difference between the realizations is the scale on the coordinate axes. The major scale change occurred between NAD 83 (1986) and NAD 83 (HARN).

NAD 83 generally does not make provision for tectonic plate movement. The fact that the North American Plate makes up the majority of the United States is convenient. The West Coast of the United States is located on the Pacific Plate and does demonstrate horizontal velocity with respect to the rest of the United States. This is a current weakness of the NAD 83 reference system.

WGS 84

WGS 84 was developed by the United States Department of Defense (now known as the National Imagery and Mapping Agency or NIMA) for mapping, charting, positioning, and navigation. The positions of the GPS satellite orbits are expressed in terms of WGS coordinates as derived from the positions of the ground tracking stations. Accordingly, absolute positioning derived from GPS satellites, using a single receiver, is expressed in WGS 84.

The first realization of WGS 84 was established in 1987 and was essentially identical to NAD 83 (1986). Both reference frames were developed using Doppler satellite observations. The WGS 84 and GRS 80 ellipsoids have identical semi-major axis lengths and slightly different flattening values. However, subsequent realizations of WGS 84 depart sharply from NAD 83 (1986) and approach more closely the ITRS reference frames.

The second realization of WGS 84, WGS 84 (G730), was established in 1994. This realization is based completely on GPS observations. The Department of Defense was not constrained to hold the approximately 2 meter error in the location of the origin or the 0.03 second misalignment of the 3-D Cartesian coordinate axes as was the NGS with subsequent realizations of NAD

83. The "G" in G730 stands for GPS and the "730" stands for the 730th GPS week number which started at 0h UTC 2 January 1994. A third realization was denoted as G873 which referred to the 873rd GPS week beginning 0h UTC 29 September 1996. The origin, orientation, and scale of these reference frames (or subsequent realizations) of WGS 84 are based on the positional coordinates of the fifteen tracking stations maintained by the U.S Air Force and NIMA.

WGS 84 is an evolving reference system and it can no longer be said that it is essentially the same as NAD 83. The decision was made to hold the Doppler satellite derived origin and orientation of the 3-D Cartesian coordinate system for all realizations of NAD 83, but the managers of WGS 84 were free to utilize the more precise measurements of advanced technologies (VLBI) for determination of the origin, orientation and scale of the coordinate axes.

ITRS

The International Earth Rotation Service established the International Terrestrial Reference System (ITRS) in 1988 for scientific work such as monitoring crustal motion and polar motion. The initial realization of the ITRS was the International Terrestrial Reference Frame of 1988 (ITRF88). Dracup reminds us of the Palmdale Bulge controversy along the San Andreas Fault in California in the 1970's where one group of scientists claimed that the ground had lifted some 10 to 16 inches near Palmdale. Another group claimed the bulge was due to instrument error. Confidence in measurements is associated with accurate reference frames.

The ITRS Center of the IERS publishes positions and velocities for several hundred control stations world wide. The ITRS almost always publishes new coordinates for these stations on a yearly basis providing many realizations or reference frames. The ITRF96 reference frame is defined by positions and velocities on 508 stations among 209 globally distributed sites. Each site may contain more than one advanced measurement technology such as GPS, VLBI, SLR, LLR (Lunar laser ranging), or DORIS (Doppler orbitology and radiopositioning integrated by satellite).

The ITRS assumes that the earth's surface as a whole does not move relative to the interior of the earth. This simplifies the model accounting for the movement of some 20 tectonic plates. These plates are assumed free to move with respect to each other so that the movement of one plate is compensated by the movement of the others. ITRF96 velocities in the continental U.S. range between 10 to 20 millimeters a year. Greater velocities are seen in Alaska and Hawaii.

Transformation from ITRS to NAD 83 (CORS96)

The ability to effect transformations between ITRF and NAD 83 has become important to the GPS user. The decision to hold the original Doppler satellite derived origin and orientation of the 3-D coordinate axes provides a significant difference between the ITRS and all realizations of NAD 83. The ITRS is used for GPS orbit calculations and for the positions of CORS stations. However, NAD 83 has been officially adopted by the federal government as the datum for surveying and mapping in the United States. It is important to realize that the NAD 83 datum offers the nationwide network of control monuments accessible to the public.

Published coordinates for ITRF00, the current reference frame for CORS stations, are provided as of 1 January 1997 with velocity coefficients to enable the computation of the position at a subsequent time. The velocity coefficients exemplify the time-dependent nature of geodetic positions resulting from plate tectonics. At another time (t) the position in x can be computed using the formula:

$$x(t) = x(1997.0) + vx * (t - 1997.0)$$

where x(t) is the x coordinate at time (t), x(1997.0) is the published coordinate on 1 January 1997, and vx is the published velocity in the x-component. Adjustments for the y and z coordinates can be accomplished similarly.

The familiar seven-parameter transformation is used to transform positions from the ITRS reference frame to the NAD 83 reference frame:

$$X_n = T_x + (1+S)X_i + R_zY_i - R_yZ_i$$
$$Y_n = T_y - R_zX_i + (1+S)Y_i + R_xZ_i$$
$$Z_n = T_z + R_yX_i - R_xY_i + (1+S)Z_i$$

where (X_i, Y_i, Z_i) are ITRS coordinates, (X_n, Y_n, Z_n) are NAD 83 coordinates, (R_x, R_y, R_z) are the rotation angles, in radians, about the X, Y, and Z axes, respectively, and S is the scale factor.

The NGS publishes tables of transformation parameters needed to transform ITRS positions into NAD 83 positions. These are currently found at www.ngs.noaa.gov/CORS/Coords.html. Also available at this site are transformation equations.

This coordinate transformation is encoded in a software package called HTDP (Horizontal Time-Dependent Positioning) and is available for transforming coordinates between ITRF00 and NAD 83 (CORS96) on the NGS Website.

Sample Problem 12.1

The published position of CORS station Dodson Butte is shown below. Using the ITRF00 (epoch 1997.0) position, compute:
      a.   the NAD 83 (CORS96) position (epoch 1997.0)
      b.   the NAD 83 (CORS) position (epoch March 28, 2002)

Antenna Reference Point(ARP): DODSON BUTTE CORS ARP   ----

          PID = AJ7211

ITRF00 POSITION (EPOCH 1997.0)
Computed in Jan., 2002 using 23 days of data.
   X = -2556649.442 m   latitude   = 43 07 07.64872 N
   Y = -3900393.567 m   longitude = 123 14 39.26348 W
   Z = 4337795.598 m   ellipsoid height = 953.176 m

ITRF00 VELOCITY
Predicted with HTDP_2.6 January 2002.
   VX = -0.0158 m/yr   northward = -0.0065 m/yr
   VY = 0.0045 m/yr   eastward = -0.0157 m/yr
   VZ = -0.0043 m/yr   upward   = 0.0006 m/yr

NAD_83 POSITION (EPOCH 1997.0)
Transformed from ITRF00 (epoch 1997.0) position in Jan., 2002.
   X = -2556648.866 m   latitude   = 43 07 07.62977 N
   Y = -3900394.783 m   longitude = 123 14 39.21270 W

Z = 4337795.452 m   ellipsoid height = 953.588   m

NAD_83 VELOCITY
Transformed from ITRF00 velocity in Jan., 2002.
   VX = 0.0023 m/yr   northward = 0.0083 m/yr
   VY = 0.0053 m/yr   eastward = -0.0010 m/yr
   VZ = 0.0060 m/yr   upward   = 0.0000 m/yr

The NGS Web site provides the following transformation equations:

$$x_{NAD83} = T_x(t) + [1+s(t)] \cdot x_{ITRF} + \varepsilon_z(t) \cdot y_{ITRF} - \varepsilon_y(t) \cdot z_{ITRF}$$

$$y_{NAD83} = T_y(t) - \varepsilon_z(t) \cdot x_{ITRF} + [1+s(t)] \cdot y_{ITRF} + \varepsilon_x(t) \cdot z_{ITRF}$$

$$z_{NAD83} = T_z(t) + \varepsilon_y(t) \cdot x_{ITRF} - \varepsilon_x(t) \cdot y_{ITRF} + [1+s(t)] \cdot z_{ITRF}$$

The NGS Web site also provides the following transformation parameters from ITRF00 to NAD 83 (CORS96):

$t_0 = 1997.0$

$T_x(t_0) = 0.9956$ m; $T_y(t_0) = -1.9013$ m; $T_z(t_0) = -0.5215$ m

$\varepsilon_x(t_0) = 25.915$ mas; $\varepsilon_y(t_0) = 9.426$ mas; $\varepsilon_z(t_0) = 11.599$ mas

$s(t_0) = 0.62 \cdot 10^{-9}$ (unitless)

$\dot{T}_x = 0.0007$ m $\cdot$ year^{-1}; $\dot{T}_y = -0.0007$ m $\cdot$ year^{-1}; $\dot{T}_z = 0.0005$ m $\cdot$ year^{-1}

$\dot{\varepsilon}_x = 0.067$ mas $\cdot$ year^{-1}; $\dot{\varepsilon}_y = -0.757$ mas $\cdot$ year^{-1}; $\dot{\varepsilon}_z = -0.051$ mas $\cdot$ year^{-1}

$\dot{s} = -0.18 \cdot 10^{-9}$ year^{-1}

where:

$$T_x(t) = T_x(t_0) + \dot{T}_x \cdot (t-t_0)$$

$$T_y(t) = T_y(t_0) + \dot{T}_y \cdot (t-t_0)$$

$$T_z(t) = T_z(t_0) + \dot{T}_z \cdot (t-t_0)$$

$$\varepsilon_x(t) = [\varepsilon_x(t_0) + \dot{\varepsilon}_x \cdot (t-t_0)] \cdot m_r$$

$$\varepsilon_y(t) = [\varepsilon_y(t_0) + \dot{\varepsilon}_y \cdot (t-t_0)] \cdot m_r$$

$$\varepsilon_z(t) = [\varepsilon_z(t_0) + \dot{\varepsilon}_z \cdot (t-t_0)] \cdot m_r$$

$$s(t) = s(t_0) + \dot{s} \cdot (t-t_0)$$

$m_r = 4.84813681 \times 10^{-9}$, conversion factor from milli-arcseconds (mas) to radians

$\varepsilon_x(t_0)$, $\varepsilon_y(t_0)$, and $\varepsilon_z(t_0)$ are differential rotations about the x_{ITRF}, y_{ITRF}, and z_{ITRF} axes respectively.

The sense of the rotations is counterclockwise (anticlockwise) positive.

a. Since we desire a NAD 83 (CORS96) position at the same epoch (1997.0) as the given ITRF00 position, all time-dependant values ($\dot{T}, \dot{\varepsilon}, \dot{s}$) are zero. Therefore,

$$T(t) = T(t_0)$$
$$\varepsilon\,(t) = \varepsilon\,(t_0)$$
$$s(t) = s(t_0)$$

$x_{NAD83} = T_x(t) + [1+s(t)] \cdot x_{ITRF} + \varepsilon_z(t) \cdot y_{ITRF} - \varepsilon_y(t) \cdot z_{ITRF}$

$x_{NAD83} = 0.9956\ m + [1 + (0.62 \cdot 10^{-9})] \cdot (-2556649.442\ m)$

  $+ [(11.599\ mas) \cdot (\,4.84813681 \times 10^{-9}\,) \cdot (-3900393.567\ m)]$

  $- [(9.426\ mas) \cdot (\,4.84813681 \times 10^{-9}\,) \cdot 4337795.598\ m]$

$x_{NAD83} = 0.9956\ m + (-2556649.444\ m) + (-0.219\ m) - 0.198\ m$

$x_{NAD83} = -2556648.866\ m$

$y_{NAD83} = T_y(t) - \varepsilon_z(t) \cdot x_{ITRF} + [1+s(t)] \cdot y_{ITRF} + \varepsilon_x(t) \cdot z_{ITRF}$

$y_{NAD83} = -1.9013\ m - [(11.599\ mas) \cdot (\,4.84813681 \times 10^{-9}\,) \cdot (-2556649.442\ m)]$

  $+ [1 + (0.62 \cdot 10^{-9})] \cdot (-3900393.567\ m)$

  $+ [(25.915\ mas) \cdot (\,4.84813681 \times 10^{-9}\,) \cdot 4337795.598\ m]$

$y_{NAD83} = -1.9013\ m - (-0.144\ m) + (-3900393.569\ m) + 0.545\ m$

$y_{NAD83} = -3900394.782\ m$

$z_{NAD83} = T_z(t) + \varepsilon_y(t) \cdot x_{ITRF} - \varepsilon_x(t) \cdot y_{ITRF} + [1+s(t)] \cdot z_{ITRF}$

$z_{NAD83} = -0.5215\ m + [(9.426\ mas) \cdot (\,4.84813681 \times 10^{-9}\,) \cdot (-2556649.442\ m)]$

  $- [(25.915\ mas) \cdot (\,4.84813681 \times 10^{-9}\,) \cdot (-3900393.567\ m)]$

  $+ [1 + (0.62 \cdot 10^{-9})] \cdot 4337795.598\ m$

$z_{NAD83} = -0.5215\ m + (-0.117\ m) - (-0.490\ m) + 4337795.601\ m$

$z_{NAD83} = 4337795.452\ m$

b. The epoch of March 28, 2002 must be expressed in decimal form. Since March 28 is the 87th day of 2002, the epoch is:
  t = 2002 + 87/365 = 2002.24
  Therefore (t – t_0) = 2002.24 – 1997.0 = 5.24 yr

$$T_x(t) = T_x(t_0) + \dot{T}_x \cdot (t-t_0) = 0.9956 \text{ m} + (0.0007 \text{ m/yr}) \cdot (5.24 \text{ yr})$$
$$T_x(t) = 0.9956 \text{ m} + 0.0037 \text{ m} = 0.9993 \text{ m}$$

$$T_y(t) = T_y(t_0) + \dot{T}_y \cdot (t-t_0) = -1.9013 \text{ m} + (-0.0007 \text{ m/yr}) \cdot (5.24 \text{ yr})$$
$$T_y(t) = -1.9013 \text{ m} - 0.0037 \text{ m} = -1.9050 \text{ m}$$

$$T_z(t) = T_z(t_0) + \dot{T}_z \cdot (t-t_0) = -0.5215 \text{ m} + (0.0005 \text{ m/yr}) \cdot (5.24 \text{ yr})$$
$$T_z(t) = -0.5215 \text{ m} + 0.0026 \text{ m} = -0.5189 \text{ m}$$

$$\varepsilon_x(t) = \varepsilon_x(t_0) + \dot{\varepsilon}_x \cdot (t-t_0) = 25.915 \text{ mas} + (0.067 \text{ mas/yr}) \cdot (5.24 \text{ yr})$$
$$\varepsilon_x(t) = 25.915 \text{ mas} + 0.351 \text{ mas} = 26.266 \text{ mas}$$

$$\varepsilon_y(t) = \varepsilon_y(t_0) + \dot{\varepsilon}_y \cdot (t-t_0) = 9.426 \text{ mas} + (-0.757 \text{ mas/yr}) \cdot (5.24 \text{ yr})$$
$$\varepsilon_y(t) = 9.426 \text{ mas} - 3.967 \text{ mas} = 5.459 \text{ mas}$$

$$\varepsilon_z(t) = \varepsilon_z(t_0) + \dot{\varepsilon}_z \cdot (t-t_0) = 11.599 \text{ mas} + (-0.051 \text{ mas/yr}) \cdot (5.24 \text{ yr})$$
$$\varepsilon_z(t) = 11.599 \text{ mas} - 0.267 \text{ mas} = 11.332 \text{ mas}$$

$$s(t) = s(t_0) + \dot{s} \cdot (t-t_0) = (0.62 \cdot 10^{-9}) + (-0.18 \cdot 10^{-9} \text{/yr}) \cdot (5.24 \text{ yr})$$
$$s(t) = (0.62 \cdot 10^{-9}) - (0.94 \cdot 10^{-9}) = (-0.32 \cdot 10^{-9})$$

$$x_{NAD83} = -2556648.771 \text{ m}$$
$$y_{NAD83} = -3900394.778 \text{ m}$$
$$z_{NAD83} = 4337795.507 \text{ m}$$

Sample Problem 12.2

Perform the coordinate transformation from Problem 12.1 using the HTDP program.

a. HTDP (version 2.6) OUTPUT

TRANSFORMING POSITIONS FROM ITRF00 (EPOCH = 01-01-1997)
            TO NAD_83 (EPOCH = 01-01-1997)

	INPUT COORDINATES	OUTPUT COORDINATES
Dodson Butte		
LATITUDE	43 07 7.64874 N	43 07 7.62979 N
LONGITUDE	123 14 39.26349 W	123 14 39.21271 W
ELLIP. HT.	953.176	953.588 m
X	-2556649.442	-2556648.866 m
Y	-3900393.567	-3900394.782 m
Z	4337795.598	4337795.452 m

b. HTDP (version 2.6) OUTPUT

TRANSFORMING POSITIONS FROM ITRF00 (EPOCH = 01-01-1997)
TO NAD_83 (EPOCH = 03-28-2002)

	INPUT COORDINATES	OUTPUT COORDINATES
Dodson Butte		
LATITUDE	43 07 7.64874 N	43 07 7.63229 N
LONGITUDE	123 14 39.26349 W	123 14 39.20931 W
ELLIP. HT.	953.176	953.585 m
X	-2556649.442	-2556648.771 m
Y	-3900393.567	-3900394.778 m
Z	4337795.598	4337795.506 m

Precise Positioning with GPS

Snay and Soler (April 2000) state "Techniques that position new points by using only satellites as control automatically yield positional coordinates that are referred to the same reference frame as that used for the orbits, either some realization of WGS 84 or some realization of ITRS."

Currently the GPS surveying technique that gives precise results is the relative positioning method. The relative positioning method involves two or more GPS receivers operating simultaneously to provide session measurements. Post-processing provides precise baselines that can be fitted between the terrestrial monuments of a reference frame; typically expressed in a realization of NAD 83.

The general lack of an accessible WGS 84 control monumentation network precludes the use of the WGS 84 reference frames for relative positioning using GPS technology. Additionally, the positional velocities of the ground tracking stations are generally unavailable to the general public.

Zilkoski, 1997 states "An important concept for the GPS user community to understand is that since precise orbits are referred to the ITRF, fixed orbit solutions of GPS produce vectors that are oriented in the ITRF coordinate system. However, most of the user community is working in NAD 83. The systematic differences between ITRF and NAD 83 must be accounted for, if high accuracy is to be maintained."

At the date of publication of this text, the NGS is using precise GPS satellite orbits (ephemeredes) referred to the International GPS Service (IGS) terrestrial reference frame of 2000 (IGS00). This reference frame is intended to be consistent with ITRF00.

We are faced with the interesting fact that satellite orbit determination is accomplished under one reference system and terrestrial project control is desired under another reference system, usually NAD 83. This underscores the importance of following the sound principles of GPS network design to impart the proper scale and orientation of the NAD 83 datum to the resulting positions. The final step of post-processing must be to perform a network adjustment constraining GPS baseline vectors to the NAD 83 Datum.

The use of CORS data will produce control coordinates in NAD 83 (CORS96] and later realizations. At the date of publication of this text, the national adjustment of the state HARNs has yet to be accomplished. The later realizations of NAD 83 can be expected to be inconsistent with local NAD 83 (1986] and NAD 83 [HARN] control. Direct mathematical transformations between the later realizations of NAD 83 and NAD 83 (1986) and NAD 83 (HARN] may not exist for your local control. While NGS provides the software program, NADCON, to handle coordinate transformations between these realizations, the program mathematically models local and regional distortions and interpolates the output. Therefore, coordinate transformations from NADCON may not be adequate for precise work.

National Adjustment of the HARN's

The development period of the state HARN's spans 1987 through 1998. The first state to implement a HARN was Tennessee, the last state was Indiana. A desirable goal is the nation-wide readjustment of all state HARNS into the National Spatial Reference System allowing rigorous coordinate transformations between the ITRF and NAD 83 (HARN). The lack of precise GPS-derived ellipsoid heights on many HARN points has frustrated the realization of this goal of a nation-wide HARN.

The implementation of the CORS stations made it possible to identify horizontal distortions and ellipsoid height distortions in existing state HARNS. The Office of National Geodetic Survey has shouldered the responsibility for the maintenance of the Federal Base Network portion of each state HARN. Those HARN's observed before 1995 may present significant upgrade efforts especially in the provision of precise ellipsoid heights.

Consider that procedures for generating precise ellipsoid height were not available until 1997. They are published in NOAA Technical Memorandum NOS NGS-58 dated November 1997 and entitled, "Guidelines for Establishing GPS-Derived Ellipsoid Heights (Standards 2 CM and 5 CM) Version 4.3."

GPS Derived Elevations

"Height modernization" is the current phrase developed to describe the intent to determine orthometric elevations by GPS surveying instead of spirit leveling. Precise control leveling using direct leveling techniques (differential leveling) requires highly trained field crews and is time consuming. Therefore, precise control leveling is expensive.

Until this time in history the three-dimensional datum was a concept that had not been realized. Horizontal datums and vertical datums developed along completely different paths. Table 12.1 demonstrates the major differences.

Table 12.1 Comparison of Horizontal versus Vertical Datums

Datum Definition	Horizontal Datums	Vertical Datums
Physical Monuments	Horizontal Control Monuments	Benchmarks
Published Coordinates	Latitude/Longitude	Elevations
Reference Surface	Ellipsoid	Mean Sea Level

One grasps the traditional gulf between horizontal and vertical datums when recovering a benchmark that was established so that it could not be occupied with a tripod. Students at Troy State University recovered a high order benchmark physically located on a railroad trestle so narrow that the railroad company had to be contacted for the train schedule before the station could be occupied by a tripod. Use of a level rod would have posed less difficulty.

This begs the question, why do you need to occupy the benchmark with a tripod? The answer is that GPS technology has finally brought the reality of a 3-D geodetic datum to fruition. This important realization was not obvious during the development of NAD 83 (1986). Few NAD 83 control stations possessed precise ellipsoid heights. Early statewide HARN development suffered the same problem. Procedures for obtaining precise elevations referenced to the ellipsoid surface had not yet been researched and published.

Such procedures have been published in NOAA Technical Memorandum NOS NGS-58 dated November 1997 and entitled, "Guidelines for Establishing GPS-Derived Ellipsoid Heights (Standards 2 CM and 5 CM) Version 4.3." The feasibility of gaining GPS-derived orthometric heights was demonstrated in the Baltimore County, Maryland NAVD 88 GPS-derived Orthometric Height Project (Henning, Carlson, and Zilkoski 1998). The student needs to be reminded that GPS Leveling can only compete with direct leveling over long distances. Over short distances the 2 to 5 centimeter error budget is unacceptable. Since the error propagation model for direct leveling assumes that error accumulates with each setup, it is possible to estimate the distance at which GPS Leveling becomes feasible. The student is referred back to the section on GPS Leveling in Chapter Nine.

Study Questions

1. List the five realizations of NAD 83 and explain the differences between each realization.
2. Why can it be said that the current realization of WGS 84 and NAD 83 (1986) are no longer almost the same?
3. What coordinate reference frames would you expect GPS satellite positions to be provided for real-time GPS applications?
4. Why can it be assumed that coordinate positions for the majority of the continental United States will have the same velocities in X, Y, and Z?
5. In what reference frame are the positions of CORS Stations, in operation since the fall of 1998, computed?
6. Explain the phrase "Height Modernization"?

7. Explain the difference between and ellipsoid height and an orthometric height? What is the reference surface for each type of elevation?
8. Explain why ellipsoid height differences and geoid height differences gave more accurate results than absolute values for ellipsoid and geoid heights in obtaining GPS-derived orthometric heights.

APPENDIX A

The NGS Data Sheet

```
See file              for more information about the datasheet.
DATABASE = Sybase ,PROGRAM = datasheet, VERSION = 6.87
1         National Geodetic Survey,   Retrieval Date = SEPTEMBER 24, 2003
 NY0977 ***********************************************************************
 NY0977  FBN           -  This is a Federal Base Network Control Station.
 NY0977  DESIGNATION -    ALTAMONT
 NY0977  PID           -  NY0977
 NY0977  STATE/COUNTY-    OR/KLAMATH
 NY0977  USGS QUAD   -    ALTAMONT (1995)
 NY0977
 NY0977                           *CURRENT SURVEY CONTROL
 NY0977
 NY0977* NAD 83(1998)-  42 12 32.56871(N)    121 44 50.17194(W)      ADJUSTED
 NY0977* NAVD 88      -       1250.3   (meters)    4102.   (feet)  GPS OBS
 NY0977
 NY0977  X            -  -2,490,031.243 (meters)                    COMP
 NY0977  Y            -  -4,024,274.156 (meters)                    COMP
 NY0977  Z            -   4,263,655.957 (meters)                    COMP
 NY0977  LAPLACE CORR-        2.97 (seconds)                   DEFLEC99
 NY0977  ELLIP HEIGHT-     1227.56 (meters)       (07/02/97) GPS OBS
 NY0977  GEOID HEIGHT-      -22.62 (meters)                   GEOID99
 NY0977
 NY0977  HORZ ORDER   -  B
 NY0977  ELLP ORDER   -  THIRD      CLASS I
 NY0977
 NY0977.The horizontal coordinates were established by GPS observations
 NY0977.and adjusted by the National Geodetic Survey in July 1997.
 NY0977.This is a SPECIAL STATUS position.  See SPECIAL STATUS under the
 NY0977.DATUM ITEM on the data sheet items page.
 NY0977
 NY0977.The orthometric height was determined by GPS observations and a
 NY0977.high-resolution geoid model.
 NY0977
 NY0977.The X, Y, and Z were computed from the position and the ellipsoidal ht.
 NY0977
 NY0977.The Laplace correction was computed from DEFLEC99 derived deflections.
 NY0977
 NY0977.The ellipsoidal height was determined by GPS observations
 NY0977.and is referenced to NAD 83.
 NY0977
 NY0977.The geoid height was determined by GEOID99.
 NY0977
 NY0977;                    North        East     Units   Scale      Converg.
 NY0977;SPC OR S    -    61,017.921 1,397,001.991  MT  1.00003361 -0 51 11.9
 NY0977;UTM  10     - 4,673,746.444   603,408.677  MT  0.99973157 +0 50 30.1
 NY0977
 NY0977                        SUPERSEDED SURVEY CONTROL
 NY0977
 NY0977  NAD 83(1991)-  42 12 32.56785(N)    121 44 50.17053(W) AD(     ) B
 NY0977  ELLIP H (02/25/91) 1227.63   (m)                       GP(     ) 4 1
 NY0977  NGVD 29 (08/24/95) 1249.16   (m)          4098.3   (f) LEVELING    3
 NY0977
 NY0977.Superseded values are not recommended for survey control.
 NY0977.NGS no longer adjusts projects to the NAD 27 or NGVD 29 datums.

 NY0977.                      to determine how the superseded data were derived.
 NY0977
 NY0977_U.S. NATIONAL GRID SPATIAL ADDRESS: 10TFM0340973746(NAD 83)
 NY0977_MARKER: DD = SURVEY DISK
 NY0977_SETTING: 7 = SET IN TOP OF CONCRETE MONUMENT
 NY0977_STAMPING: ALTAMONT
 NY0977_MARK LOGO: NGS+SS
```

```
NY0977_MAGNETIC: N = NO MAGNETIC MATERIAL
NY0977_STABILITY: C = MAY HOLD, BUT OF TYPE COMMONLY SUBJECT TO
NY0977+STABILITY: SURFACE MOTION
NY0977_SATELLITE: THE SITE LOCATION WAS REPORTED AS SUITABLE FOR
NY0977+SATELLITE: SATELLITE OBSERVATIONS - February 28, 2002
NY0977
NY0977  HISTORY      - Date      Condition        Report By
NY0977  HISTORY      - 1989      MONUMENTED       LOCSUR
NY0977  HISTORY      - 19920806  GOOD             NGS
NY0977  HISTORY      - 19941021  GOOD             NOS
NY0977  HISTORY      - 19961113  GOOD             CHANCE
NY0977  HISTORY      - 19970125  GOOD             NGS
NY0977  HISTORY      - 19971113  GOOD             NGS
NY0977  HISTORY      - 20020228  GOOD             ORTI
NY0977
NY0977                    STATION DESCRIPTION
NY0977
NY0977'DESCRIBED BY LOCAL SURVEYOR (INDIVIDUAL OR FIRM) 1989
NY0977'THE STATION IS LOCATED IN KLAMATH FALLS AT THE FAIRGROUNDS IN THE NE
NY0977'1/4 OF SECTION 3, T.39S.,R.9E.,W.M.
NY0977'TO REACH FROM THE CITY HALL IN KLAMATH FALLS PROCEED SOUTHEAST ALONG
NY0977'SOUTH FIFTH STREET AND SOUTH SIXTH STREET 3.52 KM (2.19 MI) TO ARTHUR
NY0977'STREET, TURN LEFT NAD PROCEED NORTH ON ARTHUR STREET .16 KM
NY0977'(0.10 MI) TO A PAVED ENTRANCE ROAD INTO THE KLAMATH COUNTY
NY0977'FAIRGROUNDS, TURN RIGHT AND PROCEED EAST ON THE FAIRGROUNDS ENTRANCE
NY0977'ROAD AND ITS EXTENSION APPROXIMATELY 375 FEET TO THE STATION.
NY0977'THE STATION IS A STANDARD ORDOT PRIMARY GPS DISK SET INTO A CONCRETE
NY0977'MONUMENT THAT IS 5 CM BELOW THE SURFACE OF THE GROUND.  IT IS 6.40 M
NY0977'(21.00 FT) EAST OF THE CENTERLINE OF A PAVED ROADWAY AND 8.53 M
NY0977'(27.99 FT) SOUTH OF A PAVED ROADWAY.  BOTH ROADS ARE ACCESS ROADS
NY0977'WITHIN THE KLAMATH COUNTY FAIRGROUNDS.
NY0977'DESCRIBED BY FRANCIS ROBERTS, KLAMATH COUNTY SURVEYOR.
NY0977
NY0977                    STATION RECOVERY (1992)
NY0977
NY0977'RECOVERY NOTE BY NATIONAL GEODETIC SURVEY 1992
NY0977'STATION IS LOCATED IN THE SOUTHEAST SECTION OF KLAMATH FALLS, ON THE
NY0977'WEST SIDE OF THE KLAMATH COUNTY FAIRGROUNDS, OUTSIDE THE PERIMETER
NY0977'FENCE, JUST EAST OF THE ARTHUR STREET EXIT, ABOUT 75 M (246.1 FT)
NY0977'SOUTHEAST OF THE SENIOR CITIZENS CENTER, AT JUNCTION OF PAVED
NY0977'LANES. OWNERSHIP--KLAMATH COUNTY FAIRGROUNDS, 3531 SOUTH 6TH STREET,
NY0977'KLAMATH FALLS, OR 97603.  MANAGER IS JOHN HANCOCK, PHONE
NY0977'503-883-3796.
NY0977'TO REACH FROM THE JUNCTION OF US HIGHWAY 97, STATE HIGHWAYS 66 AND 140
NY0977'ON THE SOUTHWEST SIDE OF KLAMATH FALLS, GO EAST ON HIGHWAY 140 FOR
NY0977'4.63 KM (2.88 MI) TO A PAVED CROSSROAD.  TURN LEFT, NORTH, ON
NY0977'WASHBURN WAY FOR 4.17 KM (2.59 MI) TO A CROSS STREET.  TURN RIGHT,
NY0977'SOUTHEAST, ON SOUTH 6TH STREET FOR 0.81 KM (0.50 MI) TO ARTHUR
NY0977'STREET ON THE LEFT.  TURN LEFT, NORTH, ON ARTHUR STREET FOR 0.20 KM
NY0977'(0.12 MI) TO A STREET RIGHT. TURN RIGHT, EAST, ON PAVED STREET FOR
NY0977'0.11 KM (0.07 MI) TO A CROSS LANE AND STATION STRAIGHT AHEAD.
NY0977'STATION MARK IS SET IN THE TOP OF A 30-CM ROUND CONCRETE POST 1 CM
NY0977'BELOW GROUND LEVEL IN OPEN DEPRESSION.  IT IS ON THE EXTENDED CENTER

NY0977'OF ROAD LEADING WEST TO ARTHUR STREET, 6.1 M (20.0 FT) EAST OF THE
NY0977'LANE CENTER, 8.5 M (27.9 FT) SOUTHEAST OF THE CENTER OF A PAVED
NY0977'LANE LEADING NORTHEAST TO GATE, 1.4 M (4.6 FT) SOUTHWEST OF A
NY0977'SHORT FIBERGLASS WITNESS POST AT 1 M (3.3 FT) HIGH CONCRETE POST,
NY0977'1.5 M (4.9 FT) NORTH-NORTHWEST OF A 1 M (3.3 FT) HIGH CONCRETE
NY0977'POST, AND ON THE EAST SIDE OF A ROW OF LARGE BOULDERS.
NY0977
NY0977                    STATION RECOVERY (1994)
NY0977
NY0977'RECOVERY NOTE BY NATIONAL OCEAN SERVICE 1994 (JFG)
NY0977'RECOVERED AS DESCRIBED.
NY0977
NY0977                    STATION RECOVERY (1996)
NY0977
```

NY0977'RECOVERY NOTE BY JE CHANCE AND ASSOCIATES 1996 (KAL)
NY0977'RECOVERED AS DESCRIBED BY NGS IN 1992. ALTERNATE TO REACH FOLLOWS TO
NY0977'REACH THE STATION FROM THE JUNCTION OF STATE HIGHWAY 140 AND STATE
NY0977'HIGHWAY 39 IN SOUTHEAST ALTAMONT, PROCEED WEST ON SH 39 (ALSO 6TH
NY0977'STREET) 2.7 MILES (4.3 KM) TO THE FAIRGROUNDS ENTRANCE ON THE RIGHT,
NY0977'TURN RIGHT AND PROCEED NORTH 0.2 MILES (0.3 KM) TO THE STATION ON THE
NY0977'RIGHT IN THE SOUTHEAST QUADRANT OF THE INTERSECTION OF TWO LANES
NY0977
NY0977   STATION RECOVERY (1997)
NY0977
NY0977'RECOVERY NOTE BY NATIONAL GEODETIC SURVEY 1997 (JDR)
NY0977'STATION IS LOCATED IN THE SOUTHEAST SECTION OF KLAMATH FALLS, ON THE
NY0977'WEST SIDE OF THE KLAMATH COUNTY FAIRGROUNDS, OUTSIDE THE PERIMETER
NY0977'FENCE, JUST EAST OF THE ARTHUR STREET EXIT, ABOUT 75 M (246.1 FT)
NY0977'SOUTHEAST OF THE SENIOR CITIZENS CENTER, AT THE JUNCTION OF PAVED
NY0977'LANES. OWNERSHIP--KLAMATH COUNTY FAIRGROUNDS, 3531 SOUTH 6TH STREET,
NY0977'KLAMATH FALLS, OR 97603. MANAGER IS MR JOHN HANCOCK, PHONE
NY0977'541-883-3796. TO REACH THE STATION FROM THE JUNCTION OF US HIGHWAY
NY0977'97, STATE HIGHWAYS 66 AND 140 ON THE SOUTHWEST SIDE OF KLAMATH FALLS,
NY0977'GO EASTERLY ON HIGHWAY 140 FOR 4.63 KM (2.85 MI) TO A PAVED CROSSROAD.
NY0977'TURN LEFT, NORTHERLY ON WASHBURN WAY FOR 4.17 KM (2.60 MI) TO THE
NY0977'INTERSECTION OF 6TH STREET. TURN RIGHT, SOUTHEASTERLY ON 6TH STREET
NY0977'FOR 0.81 KM (0.50 MI) TO ARTHUR STREET ON THE LEFT. TURN LEFT,
NY0977'NORTHERLY ON ARTHUR STREET FOR 0.20 KM (0.10 MI) TO A STREET ON THE
NY0977'RIGHT. TURN RIGHT, EASTERLY ON THE PAVED STREET FOR 0.11 KM (0.05 MI)
NY0977'TO A CROSS LANE AND THE STATION STRAIGHT AHEAD AND ON LINE WITH THE
NY0977'EXTENDED CENTERLINE OF ARTHUR STREET. THE STATION IS LOCATE 8.5 M
NY0977'(27.9 FT) SOUTHEAST OF THE CENTER OF A PAVED LANE LEADING NORTHEAST TO
NY0977'A GATE, 6.1 M (20.0 FT) EAST OF THE LANE CENTER, 1.4 M (4.6 FT)
NY0977'SOUTHWEST OF A SHORT FIBERGLASS WITNESS POST AT A 1.0 M (3.3 FT) HIGH
NY0977'CONCRETE POST, 1.5 M (4.9 FT) NORTH-NORTHWEST FROM A 1.0 M (3.3 FT)
NY0977'HIGH CONCRETE POST, AND ON THE EAST SIDE OF A ROW OF LARGE BOULDERS.
NY0977
NY0977   STATION RECOVERY (1997)
NY0977
NY0977'RECOVERY NOTE BY NATIONAL GEODETIC SURVEY 1997 (JDR)
NY0977'THE STATION IS LOCATED IN THE SOUTHEAST SECTION OF KLAMATH FALLS AND
NY0977'WESTERN EDGE OF ALTAMONT, ON THE WEST SIDE OF THE KLAMATH COUNTY
NY0977'FAIRGROUNDS, OUTSIDE THE PERIMETER FENCE, JUST EAST OF THE ARTHUR
NY0977'STREET EXIT FROM THE FAIRGROUNDS, ABOUT 75 M (246.1 FT) SOUTHEAST OF
NY0977'THE SENIOR CITIZENS CENTER, ON THE EAST SIDE OF A ROW OF BOULDERS IN
NY0977'THE SOUTHEAST JUNCTION OF PAVED LANES. OWNERSHIP--KLAMATH COUNTY
NY0977'FAIRGROUNDS, 3531 SOUTH 6TH STREET, KLAMATH FALLS, OR 97603. MANAGER
NY0977'IS MR JOHN HANCOCK, PHONE 541-883-3796. TO REACH THE STATION FROM THE
NY0977'JUNCTION OF U.S. HIGHWAY 97 AND STATE HIGHWAYS 66 AND 140 ON THE
NY0977'SOUTHWEST SIDE OF KLAMATH FALLS, GO EAST ON HIGHWAY 140 FOR 4.63 KM

NY0977'(2.85 MI) TO THE INTERSECTION OF WASHBURN WAY. TURN LEFT, NORTH, ON
NY0977'WASHBURN WAY FOR 4.17 KM (2.60 MI) TO THE INTERSECTION OF 6TH STREET.
NY0977'TURN RIGHT, SOUTHEAST, ON 6TH STREET FOR 0.81 KM (0.50 MI) TO ARTHUR
NY0977'STREET ON THE LEFT. TURN LEFT, NORTH, ON ARTHUR STREET FOR 0.20 KM
NY0977'(0.10 MI) TO A STREET ON THE RIGHT. TURN RIGHT, EAST, ON THE PAVED
NY0977'STREET FOR 0.11 KM (0.05 MI) TO A CROSS LANE AND THE STATION STRAIGHT
NY0977'AHEAD AND ON LINE WITH THE EXTENDED CENTER OF THE PAVED STREET. THE
NY0977'STATION IS AN OREGON PRIMARY GPS STATION DISK SET IN TOP OF A 30 CM
NY0977'ROUND CONCRETE MONUMENT FLUSH WITH THE GROUND. IT IS 8.5 M (27.9 FT)
NY0977'SOUTHEAST FROM THE CENTER OF A PAVED DRIVE LEADING NORTHEAST TO A
NY0977'GATE, 6.1 M (20.0 FT) EAST FROM THE CENTER OF THE NORTH-SOUTH LANE,
NY0977'1.4 M (4.6 FT) SOUTHWEST FROM A SHORT FIBERGLAS WITNESS POST AT A 100
NY0977'CM HIGH CONCRETE POST, AND 1.5 M (4.9 FT) NORTH-NORTHWEST FROM A 100
NY0977'CM HIGH CONCRETE POST.
NY0977
NY0977   STATION RECOVERY (2002)
NY0977
NY0977'RECOVERY NOTE BY OREGON TECHNICAL INSTITUTE 2002 (SND)
NY0977'RECOVERED AS DESCRIBED
NY0977'
NY0977'

APPENDIX B

NADCON transforms geographic coordinates between NAD 27, Old Hawaiian, Puerto Rico, or Alaska Island datums and NAD 83 values. Recommended for converting coordinate data for mapping, low-accuracy surveying, or navigation.

Sample NADCON Output

```
                    North American Datum Conversion
                            NAD 83 to NAD 27
                       NADCON Program Version 2.10


    ==============================================================


                    Transformation #:    1        Region: Conus

    Station name:  ALTAMONT

                                 Latitude              Longitude
    NAD 27 datum values:       42 12 33.04537        121 44 46.16901
    NAD 83 datum values:       42 12 32.56914        121 44 50.17624
    NAD 27 - NAD 83 shift values:    .47623               -4.00723(secs.)
                                   14.694               -91.920  (meters)
    Magnitude of total shift:                   93.087(meters)



                    North American Datum Conversion
                            HPGN   to NAD 83
                       NADCON Program Version 2.10


    ==============================================================


                    Transformation #:    1        Region:

    Station name:  ALTAMONT

                                 Latitude              Longitude
    NAD 83 datum values:       42 12 32.56914        121 44 50.17624
    HPGN datum values:         42 12 32.56785        121 44 50.17053
    NAD 83 - HPGN shift values:      .00129                .00571(secs.)
                                    .040                   .131  (meters)
    Magnitude of total shift:                   .137(meters)
```

VERTCON computes the modeled difference in orthometric height between the North American Vertical Datum of 1988 (NAVD 88) and the National Geodetic Vertical Datum of 1929 (NGVD 29) for a given location specified by latitude and longitude. This conversion is sufficient for many mapping purposes.

Sample VERTCON Output

```
                VERTical CONversion (VERTCON) Transformation Program
                            Between NGVD 29 and NAVD 88
                                    Version 1.0


========================================================================================


            Station Name:  ALTAMONT
                Latitude            Longitude      NAVD 88 - NGVD 29 (meters)
            42 12 32.56785        121 44 50.17053         1.08
```

DEFLEC99 computes deflections of the vertical at the surface of the earth for the coterminous United States, Alaska, Puerto Rico, Virgin Islands, and Hawaii.

Output from DEFLEC99

```
                            latitude        longitude      Xi       Eta     Hor Lap
        Station Name     ddd mm ss.sssss ddd mm ss.sssss arc-sec  arc-sec  arc-sec
        USER LOCATION      31 48  2.00000  85 57 26.00000   2.58    -1.57     0.97
```

GEOID99 computes hybrid geoid undulations (heights), and is used to convert between NAD 83 GPS-derived ellipsoid heights and NAVD 88 orthometric heights. Areas covered are the coterminous United States, Hawaii, Alaska, Puerto Rico and the American Virgin Islands.

Output from GEOID99

```
                                latitude        longitude       N
        Station Name         ddd mm ss.sssss ddd mm ss.sssss  meters
        USER LOCATION          31 48  2.00000  85 57 26.00000 -26.459
```

CORPSCON is a popular software program to transform coordinates between NAD 27 and NAD 83. As with NADCON, the use of CORPSCON should be limited to mapping, low-accuracy surveying, or navigation applications.

Sample CORPSCON Output

```
                    CORPSCON EXAMPLE
                    SUR 451  GEODESY                      9/9/1997

        ----------------------------------------------------------------
        Original Coordinates on NAD 83 Geographic Coordinates
        Translated Coordinates on NAD 27 Geographic Coordinates
        ----------------------------------------------------------------
        NAME                     INPUT              OUTPUT

        ALTAMONT              42 12 32.56914 N    42 12 33.04537 N
                             121 44 50.17624 W   121 44 46.16901 W
                Datum Shift(m),    Delta Lat.  = -14.694
                                   Delta Lon.  =  91.920
```

APPENDIX C

DERVIATIONS

Derivation of p and z of the meridian ellipse

Let the plane curve to be rotated about the rotation axis to form the ellipsoid of revolution be an ellipse. The equation for the ellipse in the meridian plane is:

$$\frac{p^2}{a^2} + \frac{z^2}{b^2} - 1 = 0 \; ; \qquad \text{(C.1)}$$

It is instructive to give a physical meaning to the parameters by stating the derivation in terms of the radius of the parallel circle (p) and the perpendicular distance from the equatorial plane (z).

The first derivative of z with respect to p in equation (C.1) is:

$$\frac{dz}{dp} = -\frac{p}{z}\frac{b^2}{a^2} \; . \qquad \text{(C.2)}$$

Given the meridian section of the ellipsoid of revolution, shown in figure 5.4 within the body of Chapter 5, the tangent line to the ellipse is stated as:

$$\tan(90° + \phi) = \frac{dz}{dp} \quad or \quad \tan\phi = -\frac{dp}{dz} \; ; \qquad \text{(C.3)}$$

Manipulate the formula for eccentricity:

$$e = \frac{\left(a^2 - b^2\right)^{1/2}}{a} \quad then \quad e^2 = \frac{a^2 - b^2}{a^2} = 1 - \frac{b^2}{a^2} \quad so \quad \frac{b^2}{a^2} = 1 - e^2 \; .$$

Substitute the value for b^2/a^2 in the formula for eccentricity and the value for dp/dz in equation (C.3) into equation (C.2):

$$\frac{z}{p} = (1 - e^2)\tan\phi ,$$

Square both sides of the above equation:

$$\frac{z^2}{p^2} = \left(1 - e^2\right)^2 \tan^2\phi . \qquad\qquad (C.4)$$

It is advantageous to write equation C.1 in another form:

$$p^2 + \frac{a^2 z^2}{b^2} = a^2; \text{ since } \frac{a^2}{b^2} = \frac{1}{1 - e^2} \text{ then } p^2 + \frac{z^2}{1 - e^2} = a^2;$$

$$\text{so } p^2 = a^2 - \frac{z^2}{1 - e^2}; \qquad\qquad (C.5)$$

$$\text{and } z^2 = \left(a^2 - p^2\right)\left(1 - e^2\right). \qquad\qquad (C.6)$$

Substitute the value for z^2 in equation (C.6) into equation (C.4):

$$\frac{(a^2 - p^2)(1 - e^2)}{p^2} = (1 - e^2)^2 \tan^2\phi \ or \ \frac{a^2 - p^2}{p^2} = (1 - e^2)\tan^2\phi \ ;$$

$$a^2 - p^2 = p^2(1 - e^2)\tan^2\phi \ \ or \ \ a^2 = p^2 + p^2(1 - e^2)\tan^2\phi \ ;$$

$$a^2 = p^2[1 + (1 - e^2)\tan^2\phi] \ \ and \ \ p^2 = \frac{a^2}{1 + \tan^2\phi - e^2\tan^2\phi} \ ;$$

Using the trigonometric identities:

$$1 + \tan^2\phi = \sec^2\phi \ \ and \ \ \tan^2\phi = \frac{\sin^2\phi}{\cos^2\phi} \ ;$$

$$p^2 = \frac{a^2}{\sec^2\phi - e^2 \dfrac{\sin^2\phi}{\cos^2\phi}} \ ;$$

Multiplication by $\cos^2 \phi$ *gives* $\quad p^2 = \dfrac{a^2 \cos^2 \phi}{1 - e^2 \sin^2 \phi}$; then take the square root,

$$\therefore \quad p = \frac{a \cos \phi}{\left(1 - e^2 \sin^2 \phi\right)^{1/2}} \tag{C.7}$$

Similarly to derive the formula for (z) as a function of geodetic latitude (ϕ), the value for p^2 in equation (C.5) is substituted into equation (C.4):

$$\frac{z^2}{a^2 - \dfrac{z^2}{1 - e^2}} = \left(1 - e^2\right)^2 \tan^2 \phi,$$

$$a^2 - \frac{z^2}{1 - e^2} = \frac{z^2}{\left(1 - e^2\right)^2 \tan^2 \phi},$$

$$a^2 = \frac{z^2}{1 - e^2} + \frac{z^2}{\left(1 - e^2\right)^2 \tan^2 \phi},$$

$$a^2 = z^2 \left[\frac{1}{1 - e^2} + \frac{1}{\left(1 - e^2\right)^2 \tan^2 \phi}\right],$$

$$a^2 = z^2 \left[\frac{\left(1 - e^2\right)\tan^2 \phi + 1}{\left(1 - e^2\right)^2 \tan^2 \phi}\right],$$

$$a^2 = z^2 \left[\frac{\tan^2 \phi - e^2 \tan^2 \phi + 1}{(1 - e^2)^2 \tan^2 \phi}\right].$$

Substituting in the trigonometric identity $1 + \tan^2 \phi = \sec^2 \phi$ gives:

$$a^2 = z^2 \left[\frac{\sec^2 \phi - e^2 \tan^2 \phi}{\left(1 - e^2\right)^2 \tan^2 \phi}\right],$$

so $z^2 = \dfrac{a^2 \left(1 - e^2\right)^2 \tan^2 \phi}{\sec^2 \phi - e^2 \tan^2 \phi}$,

$$z^2 = \frac{a^2 \left(1 - e^2\right)^2 \dfrac{\sin^2 \phi}{\cos^2 \phi}}{\sec^2 \phi - e^2 \dfrac{\sin^2 \phi}{\cos^2 \phi}} ,$$

Multiply through by $\cos^2\phi$;

$$z^2 = \frac{a^2(1-e^2)^2 \sin^2\phi}{1-e^2\sin^2\phi} \text{ , then take the square root,}$$

$$\therefore \quad z = \frac{a(1-e^2)\sin\phi}{(1-e^2\sin^2\phi)^{\frac{1}{2}}} . \qquad (C.8)$$

Derivation of Radius of Curvature in the Meridian Section

Given a plane curve (the meridian ellipse) whose graph or shape is adequately captured by a twice-differentiable function. The equation of the radius of curvature in terms of rectangular coordinates (z) and (p) is cited in Chapter Five and is stated as:

$$Radius\ of\ Curvature = \frac{\left(1+\left(\dfrac{dz}{dp}\right)^2\right)^{\frac{3}{2}}}{\dfrac{d^2z}{dp^2}} \qquad (C.9)$$

The first derivative of z with respect to p was given in equation (C.2). The second derivative of z with respect to p is:

$$\frac{d^2z}{dp^2} = -\frac{b^2}{a^2}\frac{d}{dp}\left[\frac{p}{z}\right] = -\frac{b^2}{a^2}\left(\frac{z-p\dfrac{dz}{dp}}{z^2}\right);$$

The first derivative is used to simply the result.

$$\frac{d^2z}{dp^2} = -\frac{b^2}{a^2z^2}\left(z+\frac{b^2p^2}{a^2z}\right) = -\frac{b^2z}{a^2z^2} - \frac{b^4p^2}{a^4z^3}.$$

244

Note: (remove square from z term in denominator within parentheses). Multiply through by z/b^2 to simplify the result:

$$\frac{d^2z}{dp^2} = -\frac{b^4 z^2}{a^2 b^2 z^3} - \frac{b^4 p^2}{a^4 z^3} = -\frac{b^4}{a^2 z^3}\left(\frac{z^2}{b^2} + \frac{p^2}{a^2}\right);$$

But this contains the original formula for the ellipse (C.1) which is equal to (1) one;

So the second derivative is $\dfrac{d^2z}{dp^2} = -\dfrac{b^4}{a^2 z^3}$. (C.10)

Placement of the first and second derivatives into Equation (C.9) gives the radius of curvature in the meridian:

$$M = -\frac{\left[1 + \dfrac{b^4 p^2}{a^4 z^2}\right]^{3/2}}{b^4\big/a^2 z^3} = -\frac{a^2 z^3 \left[a^4 z^2 + b^4 p^4\right]^{3/2}}{b^4\left(a^{12} z^6\right)^{1/2}};$$

$$M = -\frac{\left[a^4 z^2 + b^4 p^2\right]^{3/2}}{a^4 b^4}.$$ (C.11)

M is a function of the semi-major axis (a), semi-minor axis (b), radius of the parallel circle (p) and the perpendicular distance from the equator (z). It is necessary that we rearrange the equation in terms of the variables a, e, and ϕ. The semi-major axis is a defining parameter of the ellipsoid of revolution. The eccentricity (e) is computed from the defining parameters of the ellipsoid. Since (a) and (e) are constants for a given ellipsoid, the variable that changes due to position on the surface is geodetic latitude (ϕ). Therefore it can be said that the radius of curvature in the meridian is a function of latitude.

To obtain a formula for M in terms of geodetic latitude, we substitute equations (C.7) and (C.8) into equation (C.11):

$$M = \frac{\left[a^4\left(\dfrac{a\left(1-e^2\right)\sin\phi}{\left(1-e^2\sin^2\phi\right)^{1/2}}\right)^2 + b^4\left(\dfrac{a\cos\phi}{\left(1-e^2\sin^2\phi\right)^{1/2}}\right)^2\right]^{3/2}}{-b^4 a^4};$$

245

$$M = \frac{\left[a^4 \left(\dfrac{a^2 \left(1-e^2\right)^2 \sin^2 \phi}{\left(1-e^2 \sin^2 \phi\right)} \right) + b^4 \left(\dfrac{a^2 \cos^2 \phi}{\left(1-e^2 \sin^2 \phi\right)} \right) \right]^{3/2}}{-b^4 a^4} \, ,$$

$$M = -\frac{\left(a^6 (1-e^2)^2 \sin^2 \phi + a^2 b^4 \cos^2 \phi \right)^{3/2}}{b^4 a^4 \left((1-e^2) \sin^2 \phi \right)^{3/2}} \, ,$$

Since $\dfrac{b^2}{a^2} = 1 - e^2$ then $b^2 = a^2 \left(1 - e^2\right)$,

and $M = -\dfrac{\left(a^6 \left(1-e^2\right)^2 \sin^2 \phi + a^2 \left(a^2 \left(1-e^2\right) \right)^2 \cos^2 \phi \right)^{3/2}}{a^4 \left(a^2 \left(1-e^2\right) \right)^2 \left(1-e^2 \sin^2 \phi\right)^{3/2}}$,

$$M = -\frac{\left(a^6 \left(1-e^2\right)^2 \left(\sin^2 \phi + \cos^2 \phi \right) \right)^{3/2}}{a^8 \left(1-e^2\right)^2 \left(1-e^2 \sin^2 \phi\right)^{3/2}} \, ,$$

$$M = -\frac{a^9 \left(1-e^2\right)^3}{a^8 \left(1-e^2\right)^2 \left(1-e^2 \sin^2 \phi\right)^{3/2}} \, .$$

The minus sign is dropped since it indicates only the direction of bending.

$$\therefore \quad M = \frac{a\left(1-e^2\right)}{\left(1-e^2 \sin^2 \phi\right)^{3/2}} \, . \qquad\qquad (C.12)$$

BIBLIOGRAPHY

Anderson, James M., and Mikhail, Edward M. *Surveying: Theory and Practice.* New York: WCB/McGraw-Hill, 1998.

Bomford, G. *Geodesy.* 3rd Edition. Oxford: Oxford University Press, 1971.

Bowring, B.R. "The Direct and Inverse Solutions for Short Geodesic Lines on the Ellipsoid." Surveying and Mapping, 41(2): 135-141.

Bureau of Land Management. *Manual of Instructions for the Survey of the Public Lands of the United States.* United States Department of the Interior, 1973.

Burkholder, Earl F. "Computation of Horizontal/Level Distances." Journal of Surveying Engineering, 117(3): 104-116.

Burkholder, Earl F. "Geodesy." Chapter 12 of *The Surveying Handbook.* New York: Van Nostrand Reinhold, 1987.

Dracup, Joseph F. "Geodetic Surveying 1940-1990. http://www.ngs.noaa.gov:80/geodetic_surveying_1940.html.

Dutton, Benjamin. *Navigation and Nautical Astronomy.* 8th Edition. Annapolis: United States Naval Institute, 1943.

Ewing, Clair F., and Mitchell, Michael M. *Introduction to Geodesy.* New York: Elsevier Scientific Publishing Co., 1970.

Federal Geodetic Control Committee. "Geometric Geodetic **Accuracy** Standards and Specifications for Using GPS Relative Positioning Techniques." Draft version 5.0, reprinted with corrections. 1989.

Federal Geodetic Control Committee. "Standards and Specifications for Geodetic Control Networks." 1984.

Glossary of the Mapping Sciences. Published by American Society of Civil Engineers, American Congress on Surveying and Mapping, and American Society for Photogrammetry and Remote Sensing, 1994.

Heiskanen, W. "On the World Geodetic System." Publications of the Institute of Geodesy, Photogrammetry and Cartography No. 1, The Graduate School, The Ohio State University, Columbus, Ohio. 1951.

Heiskanen, W.A., and Meinesz, F.A. Vening. *The Earth and its Gravity Field.* New York: McGraw-Hill, 1958.

Heiskanen, W.A., and Moritz, H. *Physical Geodesy.* San Francisco: W.H. Freeman, 1967.

Henning, William E., Carlson, Edward E., and Zilkoski, David B. "Baltimore County, Maryland, NAVD88 GPS-derived Orthometric Height Project." Surveying and Land Information Systems, Vol 58, No. 2, 1998, pp 97-113.

International Earth Rotation Service. Various Web pages. http://www.iers.org/iers/earth.

James and James. *Mathematics Dictionary.* Edited by James, Glen. Fourth Edition. New York: Van Nostrand Reinhold Company, 1976.

Kissam, Phillip. *Surveying for Civil Engineers.* New York: McGraw-Hill Book Company, 1956.

Larson, Ron., Hostetler, Robert P., and Edwards, Bruce H. *Calculus with Analytic Geometry.* Seventh Edition. Houghton Mifflin Company, 2002. p. 825.

Leick, Alfred. *GPS Satellite Surveying.* 2nd Edition. New York: John Wiley and Sons, Inc.,1995.

Love, John D., and Strange, W.E. "High Accuracy Reference Networks: A National Perspective." Presented at the ASCE Specialty Conference- "Transportation Applications of GPS Positioning Strategy." Sacramento, CA, September 18-21, 1991.

Ma, C., et.al. "The International Celestial Reference Frame as Realized by Very Long Baseline Interferometry." The Astronomical Journal, 116: 516-546.

Mackie, J.B. *The Elements of Astronomy for Surveyors.* Charles Griffin and Company Ltd., 1985.

Maling, D.H. *Coordinate Systems and Map Projections.* 2nd Edition. New York: Pergamon Press, 1992.

McEntyre, John G. *Land Survey Systems.* Rancho Cordova: Landmark Enterprises, 1985.

Mikhail, Edward M. *Observations and Least Squares.* New York: IEP—A Dun-Donnelly Publisher, 1976.

Moritz, H. "The Definition of a Geodetic Datum." Second International Symposium on Redefinition of North American Networks." Arlington, Virginia, April 24-28,

National Geodetic Survey. *Geodetic Glossary.* National Geodetic Information Branch, NOAA, Rockville, MD, 1986.

National Geodetic Survey. "Geodetic Surveying." NOAA Manual NOS NGS 3. Rockville , Maryland 1981.

National Geodetic Survey. *North American Datum of 1983.* NOAA Professional Paper NOS 2. Edited by Charles R. Swartz. U.S. Department of Commerce. December 1989.

National Geodetic Survey. Readme file for DEFLEC99. http://www.ngs.noaa.gov/GEOID/DEFLEC99/x1999rme.txt, January 12, 2000.

National Geodetic Survey. Readme file for GEOID99. http://www.ngs.noaa.gov/GEOID/GEOID99/ January 12, 2000.

National Geodetic Survey. "Use of Calibration Base Lines." NOAA Technical Memorandum NOS NGS-10. Rockville, MD., December 1977.

National Imagery and Mapping Agency. "Department of Defense World Geodetic System 1984." Technical Report 8350.2, 3rd Edition, Amendment 1, January 3, 2000.

National Oceanic and Atmospheric Administration. "Geodesy for the Layman." Fifth Edition. July 1985.

NOAA Technical Memorandum NOS NGS-58. Guidelines for Establishing GPS-Derived Ellipsoid Heights (Standards: 2 CM and 5 CM) Version 4.3. November 1997.

Rapp, Richard H. "Geometric Geodesy Part I". (Course Notes). Department of Geodetic Science, The Ohio State University, Columbus, Ohio. April 1991.

Robillard, Walter G. *Instructions to U.S. Deputy Surveyors in Arkansas-Mississippi-Louisiana-Florida 1811-1844.* Atlanta: U.S. Department of Agriculture/Forest Service, 1970

Seeber, Günter. *Satellite Geodesy.* Berlin: Walter de Gruyter, 1993.

Sobel, Dava. Longitude: *The True Story of a Lone Genius Who Solved the Greatest Scientific Problem of His Time.* New York: Penguin Books, 1996.

"Soviet Satellite Sends U.S. Into a Tizzy." Life Magazine. Vol. 43, No. 16, October 14, 1957.

Snay, Richard A., and Soler, Thomas. "Modern Terrestrial Reference Systems (Part 1)." Professional Surveyor, December 1999.

Snay, Richard A., and Soler, Thomas. "Modern Terrestrial Reference Systems (Part 2) The Evolution of NAD 83." Professional Surveyor, February 2000.

Snay, Richard A., and Soler, Thomas. "Modern Terrestrial Reference Systems (Part 3) WGS 84 and ITRS." Professional Surveyor, March 2000.

Snay, Richard A., and Soler, Thomas. "Modern Terrestrial Reference Systems (Part 4). Practical Considerations for Accurate Positioning" Professional Surveyor, April 2000.

Spofford, Paul R. "GPS CORS and Precise Orbit Data from the National Geodetic Survey" National Geodetic Survey. Undated.

Stewart, Lowell O. *Public Land Surveys.* Minneapolis: The Meyers Printing Co., 1975.

Strang, Gilbert and Borre, Kai. *Linear Algebra, Geodesy, and GPS.* Wellesley-Cambridge Press, 1997.

Strange, William E. "Do You Really Have WGS 84 Coordinates?" Professional Surveyor, October 2000.

Strange, William E. "A National Spatial Data System Framework: Continuously Operating GPS Reference Stations. National Geodetic Survey. Undated.

Torge, Wolfgang. *Geodesy.* 2nd Edition. Berlin: Walter de Gruyter, 1991.

Twain, Mark. *Life on the Mississippi.* Reprint edition. Modern Library, 1994.

United States Code, Title 15-Commerce and Trade, Chapter 6-Weights and Measures and Standard Time., Subchapter IX-Standard Time.

Vaníček, P., and Krakiwsky, E. *Geodesy: The Concepts.* Amsterdam: North-Holland, 1982.

Wahl, Jerry L., Hintz, Raymond J., and Parker, Blair. "Geodetic Aspects of Land Boundaries and the PLSS Datum in a Cadastral Computation System." http://www.cadastral.com/cad-datm.htm, July 1996.

Zilkoski, David B., Richards, John H., and Young, Gary M. "Results of the General Adjustment of the North American Vertical Datum of 1988." Surveying and Land Information Systems, Volume 52, Number 3, 1992, pp. 133-149.

Zilkoski, David B., D'Onofrio, Joseph D., and Frakes, Stephen J. "Guidelines for Establishing GPS-Derived Ellipsoid Heights (Standards: 2 CM and 5 CM)" Version 4.3. NOAA Technical Memorandum NOS NGS-58, November 1997.

INDEX

Names of the authors and sources cited in the text are not included in the index.

251